# The Changing University

## How Increased Demand for Scientists and Technology is Transforming Academic Institutions Internationally

# NATO ASI Series

**Advanced Science Institutes Series**

*A Series presenting the results of activities sponsored by the NATO Science Committee, which aims at the dissemination of advanced scientific and technological knowledge, with a view to strengthening links between scientific communities.*

The Series is published by an international board of publishers in conjunction with the NATO Scientific Affairs Division

| | |
|---|---|
| **A Life Sciences**<br>**B Physics** | Plenum Publishing Corporation<br>London and New York |
| **C Mathematical and Physical Sciences**<br>**D Behavioural and Social Sciences**<br>**E Applied Sciences** | Kluwer Academic Publishers<br>Dordrecht, Boston and London |
| **F Computer and Systems Sciences**<br>**G Ecological Sciences**<br>**H Cell Biology**<br>**I Global Environmental Change** | Springer-Verlag<br>Berlin, Heidelberg, New York, London, Paris and Tokyo |

**NATO-PCO-DATA BASE**

The electronic index to the NATO ASI Series provides full bibliographical references (with keywords and/or abstracts) to more than 30000 contributions from international scientists published in all sections of the NATO ASI Series.
Access to the NATO-PCO-DATA BASE is possible in two ways:

– via online FILE 128 (NATO-PCO-DATA BASE) hosted by ESRIN, Via Galileo Galilei, I-00044 Frascati, Italy.

– via CD-ROM "NATO-PCO-DATA BASE" with user-friendly retrieval software in English, French and German (© WTV GmbH and DATAWARE Technologies Inc. 1989).

The CD-ROM can be ordered through any member of the Board of Publishers or through NATO-PCO, Overijse, Belgium.

**Series D: Behavioural and Social Sciences - Vol. 59**

# The Changing University

## How Increased Demand for Scientists and Technology is Transforming Academic Institutions Internationally

edited by

Dorothy S. Zinberg

Center for Science and International Affairs,
John F. Kennedy School of Government, Harvard University,
Cambridge, Massachusetts, U.S.A.

**Kluwer Academic Publishers**

Dordrecht / Boston / London

Published in cooperation with NATO Scientific Affairs Division

Proceedings of the NATO Advanced Research Workshop on
The Changing University and the Education and
Employment of Scientists and Engineers
Cambridge, Massachusetts, U.S.A.
8–10 March, 1990

ISBN 0-7923-1281-3

---

Published by Kluwer Academic Publishers,
P.O. Box 17, 3300 AA Dordrecht, The Netherlands.

Kluwer Academic Publishers incorporates the publishing programmes of
D. Reidel, Martinus Nijhoff, Dr W. Junk and MTP Press.

Sold and distributed in the U.S.A. and Canada
by Kluwer Academic Publishers,
101 Philip Drive, Norwell, MA 02061, U.S.A.

In all other countries, sold and distributed
by Kluwer Academic Publishers Group,
P.O. Box 322, 3300 AH Dordrecht, The Netherlands.

*Printed on acid-free paper*

---

## CONTENTS

## SECTION II: CURRENT TRENDS IN THE EDUCATION OF S&ES



## SECTION III: THE FLOW OF INFORMATION

## PREFACE

This collection of papers was written for a NATO Advanced Research Workshop held at Harvard University in March 1990. The title, "The Changing University and the Education of Scientists and Engineers: An International Workshop," broad as it is, does not convey the sweep of data, information, opinions, and suggestions for future research and policy choices that were crowded into two-and-a-half days of formal presentations, mealtime discussions, and teatime chats.

The proposal for the workshop grew out of a research project I had carried out that explored the policies governing the education of foreign science and engineering students (S&Es) in several industrialized countries, and of two countries that send large numbers of S&E students abroad – the People's Republic of China (PRC) and Japan (see chapter 7). In research visits to these countries as well as to France, the United Kingdom, West Germany, and within the United States, I was struck by the similarity of issues that were raised. One was the concern that there would not be enough well-trained scientists and engineers to meet the constantly increasing demand for them. Government officials, industrialists, and educators repeatedly stressed that a well-educated and -motivated work force was essential for their economies, national security, and for society as a whole. Many of those interviewed mentioned that universities are undergoing rapid, systemic changes as governments and industry are calling on them to provide human resources and intellectual capital. Yet the many government, industrial, and institutional policies effecting these changes are uncoordinated and often conflict with one another. In addition, the participants in the foreign-student study repeatedly referred to the rapid internationalization of S&Es. Markets for people, knowledge, capital, and products were no longer modified by "world" or "international," but by the vaster image of "global." All this activity is bringing about changes not yet fully comprehended, certainly not understood cross-nationally.

Furthermore, many of the people I interviewed did not know of related work going on in other countries or even within their own countries because the subject of human resources and the changing university is not a discipline but rather a loose congerie of related topics. Economists, demographers, social psychologists, admissions officers, university presidents, foreign-student recruiters, professors, government foreign-affairs and education officials, and industrialists are not likely to read the same journals or meet professionally very often. Yet the majority of participants in the study emphasized how important it was to understand what was happening with the education of S&E students within the context of the changing university and the diverse array of social-political institutions within which the university is embedded.

Organizing a workshop to bring together professionals with similar concerns, if dissimilar ways of perceiving them, was the logical next step. The different perspectives would help participants gain new insights into their own work and illuminate the interconnections. One of the purposes of the workshop was to help create an international network of individuals working on different aspects of S&E policy. I anticipated that the different perspectives would help each of us gain new insights into our own work and illuminate the interconnections. We would aim to establish a network of people with diverse backgrounds and professions who could contribute to understanding how universities are changing both internally and in their relationships with government, industry, and the public – and also serve as links to their primary groups. Clearly, new policies are needed to maintain the traditional strengths of universities as they adapt to changes in the larger world. Simply chronicling these changes nationally and internationally and conveying their significance to a larger audience could be a solid step toward the definition of critical issues, proposals for research projects, and the formulation of future policies.

The participants included some 40 individuals, many with several professional identities – politician/academic, scientist/administrator, social scientist/science-policy professional, lawyer/administrator, engineer/industrialist, members of the NATO Science Council, the majority of whom are practicing scientists, and science-policy graduate students from several countries.

Because of the diversity of topics, I decided not to publish the panel presentations but to request that those who wrote papers for the workshop incorporate the new ideas and suggestions derived from the panels and audience discussions. Unfortunately, a number of invitees from outside the U.S. were unable to attend because of schedule conflicts, which accounts for the imbalance of U.S.-focused papers.

Different languages, writing styles, vocabularies, acronyms, and discipline-bound neologisms would have been daunting had it not been for the editorial skill with which Kimberly French wielded a very sharp pencil. Her repeated question, "What does that mean?" forced us to rethink what we thought was self-explanatory. Editor Miriam Avins stepped in to prepare the final manuscript for the publisher, no easy task in this era of digitized manuscripts.

Core funding for the meetings was generously provided by the NATO Scientific Affairs Division and the MacArthur Foundation. Additional funding was made available by the Sloan Foundation, IBM, and the Social Science Research Council. A previous grant from the American Council on Germany made it possible to include a West German researcher. Funding from the National Science Foundation and the MITRE Corporation laid the groundwork during the past three years for the research leading to the workshop, which we hope will become an ongoing project. The group ended the meeting with plans to meet within the next 18 months in the center of the new Europe, Berlin.

Throughout the long, complicated arrangements for the workshop the staff of the Program for Science, Technology, and Public Policy (STPP) – administrative director Susan Fox along with staff assistants Graceann Todaro and Nora Hickey – assumed enormous responsibility and carried it out with remarkable professional skill and unfailing good humor. Tom Parris, a Kennedy School student well versed in the potential of the computer to aid in organizing and revising plans and programs deftly managed the organizational details. My colleagues Professors Lewis M. Branscomb, Harvey Brooks, Ashton B. Carter, Paul Doty, and Dr. Gerald L. Epstein contributed to the many revisions of topics, participating actively at every stage of planning, and were gamely forbearing when I was unavailable for much of anything else.

All of us at STPP hope this collection of papers will serve as a launching pad for new ideas and increased attention by everyone interested in the future of science and engineering and the university. Documenting the changes that are occurring will enable an even larger network of concerned individuals and institutions to strengthen what worked in the past and to identify the opportunities and pitfalls along the way toward the growing interdependence of S&Es, universities, industry, and government worldwide.

# INTRODUCTION

DOROTHY S. ZINBERG

"I name technology, youth exchange, universities, science, and culture."[1] So began German chancellor Helmut Kohl's ringing declaration at his historic meeting with Soviet president Mikhail Gorbachev in July 1990. Kohl's list of the building blocks on which to erect Europe's future constituted a crucial part of the two leaders' nonaggression pact. In quick order they arranged to increase the numbers of exchange students between Germany and the USSR and to encourage collaborative research between their universities. Clearly Kohl and Gorbachev believe education, human resources, and science and technology have great significance for the future of their nations and international relations. Kohl's strong words underscored the vital roles their governments' expect universities to fulfill in the achievement of peace and prosperity, and vividly demonstrated how elevated universities have become in global politics.

Those are the same themes that an international group of science-policy experts discussed a few months before at "The Changing University and the Education of Scientists and Engineers: An International Workshop," a NATO Advanced Research Workshop convened by the Program for Science, Technology, and Public Policy at Harvard University's Kennedy School of Government, March 8-10, 1990. As the papers for the workshop show, not only government but also universities, industry, and the public hold high expectations for science and engineering (S&E) education and university-based research to contribute to national, international, and global welfare.

## Background

Since the end of World War II, particularly with the recent rapid growth of the information sciences, science and technology have been acknowledged as crucial for national military and economic security. As a result, science and technology policy has become a concern at the highest levels of government. Presidents and prime ministers are likely to bring their science advisers with them to international meetings to provide guidance on collaborative projects or to scrutinize proposals for joint military endeavors that involve technology a nation would prefer to keep at home for its civilian economy. The dramatic lifecycle of the U.S. Strategic Defense Initiative (SDI), a project that

[1] Quoted by Craig Whitney in "Kohl Outlines a Vision: A Neighborly Germany," the *New York Times*, July 18, 1990, p.A6; originally reported in "News Conference in Zheleznvodsk," *Pravda*, July 16, 1990.



*D. S. Zinberg (ed.), The Changing University*, 1–14.

promised both military and economic gains, highlighted the importance of science policy for governments. The project also produced unanticipated results. Two years following SDI's inception the French, convinced that SDI was a smokescreen for U.S. government support to industry, launched their own initiative to counter the perceived technological challenge from the American SDI program. Their successful venture, EUREKA, currently involves 19 European countries and funds more than 400 projects combine basic research and development (R&D) with military and commercial aims; numerous scientists and engineers (S&Es) signed petitions aimed at banning SDI funds in their respective universities. The change in the political climate brought about by *glasnost* in the USSR and the decimation of the SDI budget resolved much of the controversy, but the close coupling of foreign policy to science and technology policy remains evident in the stronger science-advisory apparatus that has evolved in the majority of NATO countries in the past decade.

Just as foreign policy can have an impact on domestic policy, the reverse is also true. In Germany, for example, despite the availability of adequate government funding and scientific expertise, memories of the pre-1945 eugenics movement to breed a pure Aryan stock, combined with a strong contemporary ecological movement, has slowed the pace of biotechnology research. In this instance, domestic policy has repercussions for foreign policy, as well as for university research and the education of S&Es.[2]

A more specific example of how domestic academic policy can affect foreign policy is the United Kingdom's scheme to increase revenues for higher education in the early 1980s. At that time the government began to require foreign students (except for European Community nationals, refugees, and reciprocal-exchange students) to pay full fees, an increase of 300 percent. The numbers of foreign applicants plummeted. In S&E departments, which had increased their foreign-student enrollments by 150 percent between 1970 and 1978, foreign entrants fell by 50 percent. In addition, disgruntled governments unleashed a furious backlash. Malaysia, for example, launched a "Buy British Last" campaign that cost British companies an estimated 500 million pounds in lost sales. The British foreign office intervened and established scholarships for foreign applicants from its budget, thereby relieving the pressure on universities and colleges. By making the policy one of high-level government action, the U.K. was able to reverse the effects of an ill-considered move that had strained diplomatic and economic ties with the countries whose students had been adversely affected.

The foundation for a country's success in science and technology is, of course, education, which has risen in importance in industrialized as well as emerging nations. No matter how well endowed with natural resources, a country's ability to secure its position in the world order will be uncertain at best without a well-educated and -motivated work force. Much of the science and technology education of the future will

---

[2] Alan D.Beycheron, "Trends in the Twentieth Century German Research Enterprise," *The Academic Research Enterprise within Industrialized Nations: Comparative Perspectives* (National Academy of Sciences, 1990).

take place outside universities – in technical training programs, community colleges, specialized schools, and increasingly in industry. Nevertheless, the greatest expectations are of universities, the recipients of the lion's share of research funding.

Great expectations can exert great stress. If basic science in the university is expected to lead directly to discernible health benefits, information sciences to national economic competitiveness, or condensed-matter physics to military security, the real purposes of the university – namely teaching and research in an open environment – can be severely undermined.

Ironically, the accomplishments of universities around the world in the decades following World War II have been so far-reaching that the very characteristics and practices that made them productive – autonomy and openness in the exchange of information and scholars – are now threatened. The appropriateness of military and industrial funding for universities, particularly in the United States, has become even more controversial as competition for funds becomes more intense. The relatively new presence of large foreign investment in universities, such as endowed chairs or sponsored research projects, sets off alarms in quarters that fear a loss of intellectual property and potential economic strength; others fear the university's commercialization is likely to make it industry and government's handmaiden. Yet others argue that universities that make links with industry are rightfully responding to the greater need for well-trained S&Es, new knowledge, and applied science.

One example of the issues in these debates made front-page headlines in the *International Herald-Tribune* in late 1990. Critics both from within and outside the Massachusetts Institute of Technology blasted a $10-million deal it made with the Japanese. The institute agreed to duplicate a media lab that encourages young scientists to merge computers, television, and film to develop futuristic products, such as high-definition and interactive television. Some fear the contract gives away the only edge the U.S. has in its economic race with Japan – creativity. Yet MIT lab researchers argue the more labs and scientists working on such ideas, the better for the university, science, and the world.[3] This is but one of the struggles universities face as they try to balance their national identities and funding with the pulls toward internationalism.

While participants at the NATO workshop identified these and many other tensions, they also acknowledged the remarkable strides that had been made in many countries' university research programs in recent decades. The Europeans enthusiastically reported the mobility of students and faculty and the increased number of collaborative research efforts in scientific R&D across the continent. In 1990 more than 1 million students were studying engineering in Western Europe, plus an additional 400,000 in Central and Eastern Europe.[4] A striking feature of the European presentations was their uncharacteristic optimism about an increasingly mobile, intra-European S&E work force that will further

---

[3] Gina Kolata (New York Times Service), the *International Herald-Tribune*, December 20, 1990, p. 1.

[4] W.J. McGregor Tegart, quoted in an unpublished report to the Australian Science and Technology Council, June 1990.

hasten economic resurgence. Even the British, who are dismayed about the nearly lethal cuts former prime minister Margaret Thatcher's government imposed on their universities, envision a more robust future because of the influx of European S&E students and the increase in EC-funded projects and new industrial consortia.

The Japanese participants cite successful efforts to increase the percentage of students attending universities, up from about two percent of the 18-year-old population before World War II to the present 37 percent. In addition, the Japanese are attracting a larger share of students from Pacific Rim countries as part of their goal to become the educational leaders in S&E in Asia and to further their plans to internationalize.

Nevertheless, the predominant mood at the workshop was one of angst. Recent data on budget cuts, increased overhead, declining birthrates in industrialized countries, and inadequate supplies of young S&Es in some countries, coupled with high societal expectations, underlay the conviction of the participants that universities, colleges, and technical schools must take action. Since the workshop, British science has suffered further severe budget cuts. The *London Times* reports a "deepening crisis in British science and funding" as the Science and Engineering Research Council faces a 40 million pound shortfall and the research of entire departments in fields such as X-ray astronomy have been judged expendable.[5] In addition, the Medical Research Council, the leading provider of funds for biomedical research in the U.K., closed three units and canceled all funding for six months. The edict read: "No new research support, no hiring, and no new equipment."[6]

In the U.S., Leon Lederman, a Nobel Prize winner in physics and president-elect of the American Association for the Advancement of Science (AAAS), released the results from a survey of 250 researchers at 50 U.S. universities. Titled "Science: End of the Frontier?" the report "paint[s] a picture of an academic research community beset by flagging morale, diminishing expectations, and constricting horizons."[7] On a scale of one to 10, the average morale level is 2.5, Lederman estimates, because scientists are spending more time writing grant proposals yet are experiencing great difficulty funding research projects.[8]

## Perspectives from Academia, Industry, and Government

Although few participants were as gloomy as those in Lederman's grim report, an anxious tone characterized the speakers' presentations. The welcoming remarks of Harvard University president Derek C. Bok provide an insider's perspective of the strengths and

---

[5] The London Times, December 22, 1990, p. 1.

[6] Constance Holden, *Briefing in Science* 250 (November 1990): 1,202.

[7] Hilts, Philip J., "Nobel Physicist Raises Alarm on U.S. Science," the *New York Times*, National Edition, January 6, 1991, p. 16.

[8] Kevin Pope, "Funding Problems Demoralize U.S. Scientists," *The Sciences* (January/February 1991).

limitations of the university. While the university's growing involvement with industry and government provides opportunity, he notes, this change also brings difficult dilemmas. On the one hand, government and industry seek out university scientists for consulting, and some scientists have established their own commercial ventures. On the other hand, this activity diminishes the time spent teaching and carrying out research, and the new arrangements raise complex questions about patents, intellectual property, and conflict of interest. Most pointedly Bok emphasizes that universities' link to industry as a way to reinvigorate the economy has been overvalued. "If we try to justify university research and its linkage to industry as a decisive factor in economic competitiveness and try to exploit competitiveness as a reason for building science and links with industry," he admonishes, "we are likely to end by being disappointing to our constituents to our ultimate disadvantage." Other negative consequences could follow, he says, such as excessive secrecy and nationalistic policies that would hinder foreign students and companies from access to U.S. universities. He has been urging Harvard to increase its overseas-student population, recruit foreign faculty, and enhance the curriculum to adapt to a growing globalism. The university graduate, he argues, has to be prepared for a different world from that of earlier generations, so the university must change radically to make this possible. Much of what is expected of universities is contradictory, he concludes, and universities now have to reconcile those contradictions.

John Armstrong, the chief scientist of IBM (USA), takes issue with Bok. His perspective as a representative of industry is more nationalistic. Armstrong argues that the university should have its nation's economic well-being at the core of its goals and insists that universities, government, and industry resolve their conflicting views of the role of foreign-based corporations in universities, intellectual-property rights, and the conflicts of interest that arise when scientists accept public funds for their research and then exploit their inventions in the marketplace. In his chapter, "An Industry Perspective on the Changing University," Armstrong argues, "It would be a mistake to ignore the substantial lack of sympathy, both in corporations and in government, for the ethos of the open, internationally minded university community." He anticipates increasing friction between S&E faculty and administrators as well as between universities and the larger society that supports them. Although he advocates thoughtful discussion, his remarks leave little doubt that policy consensus is not likely to be achieved in the near future.

Many members of the U.S. Congress, such as Representative Ted Weiss, a Democrat from New York, have publicly aired the controversies Armstrong raises, especially about conflicts of interest in the use of public funding. Many other countries' government officials also expect universities to contribute to national economic well-being. For example, in 1987 the People's Republic of China ruled that all scientific institutes, even in the field of theoretical physics, must prove a relevance to a commercial product. And in the U.K. in 1987 the University Grants Committee, chaired by a mathematics professor, was replaced by the University Finance Committee, chaired by an engineer who was formerly the director of a British corporation. The message was not lost on the British academic community. These trends are evident in the majority of industrialized countries and contribute to the underlying tone of dismay among the participants.

In the chapter "Positioning U.S. Science Policy for the New World Order," Eric Bloch, who at the time of the workshop was the director of the National Science Foundation (NSF), strongly supports tight links between U.S. industry and academic research. Yet he believes that insisting on tangible, profitable results could endanger basic research projects. Acknowledging that U.S. success in academic research has been based on a well-educated cadre of S&Es – many of whom were educated by the G.I. Bill after World War II or were European professional scientists who fled from the Nazis – Bloch argues that a way has to be found to make careers in science more attractive to U.S. citizens. As world competition increases, the educated work force, science research programs, and the ability to communicate results rapidly will determine world economic leadership. He anticipates that the shift of federal research funds from the Department of Defense to the NSF and the National Institutes of Health will continue. However, he admits, "a continual gap will exist between the increases [in funds] and what is needed." And, he cautions, a growing world economic downturn could cause governments to adopt protectionist policies that would interfere with the communication of scientific research results, cooperation, and even the education of S&Es – the strategic resources in today's global economy.

Taking a long view of the history of the university, George Bugliarello, president of Polytechnic University (U.S.), in "The University – and Particularly the Technological University: Pragmatism and Beyond," points out that the university is an institution that has survived by avoiding too sharp a definition of what it is and what it does. He cautions that the time has come to assess what universities can and cannot do. In fact, "it is not always clear whether the great range of activities in which many universities are engaged today represent a deep ideological commitment or simply a manifestation of the need to survive." The world is rapidly becoming globalized, yet universities remain essentially national institutions; not one is supported by the global community. Universities, he adds, carry out teaching and research well, but they do not know how to attack major social problems. The technological university, however, is uniquely capable of responding to some of these problems – nationally and internationally – because it can develop close relationships with industry, he argues.

His own university recently opened a 16-acre university-industry park, Metrotech, in New York City, that will focus on telecommunications and information technologies, the bedrock of the city's financial economy. Metrotech will draw on faculty expertise, generate income for the university, provide jobs for some 15,000 people, and will train approximately 35,000 employees a year. In addition, by linking up with foreign industry, as it has with an Italian communications conglomerate, Metrotech will educate an international cadre of engineers, which will help thrust the university into a more global environment. Bugliarello believes ventures like this, already cropping up in other parts of the world – Technopark, a joint Soviet-Italian center is a noteworthy similar venture – will help the university redefine its purpose and, in so doing, assure its survival.

The politician in the group was less sanguine. In "The Swing of the Pendulum: Financing of British Universities from the 1960s through the 1980s," Shirley Williams, former Labour Party MP in the U.K. House of Commons and former Secretary of State for Education and Science, laments scientists' loss of freedom in determining which university projects will be funded. In the 1960s, she notes, "[t]he interests of industry and commerce went virtually unrepresented . . . [and] few scientists in the U.K. at that time thought of the commercial value of their discoveries." Peer review and block grants – that is, grants that went to institutions rather than to specific projects – allowed "interest-led" research to burgeon. This easy relationship between the scientist and the funding organization began to fade by the late 1960s, as universities expanded rapidly. The promised link between basic research and economic growth did not materialize, and the costs of doing research spiraled. The public came to regard science and technology with skepticism due to problems with nuclear power plants and weapons, pesticides, and inadequately tested pharmaceuticals, despite other benefits to society. Scientists themselves, once quiet about their disagreements over issues in the field, become more public with them. The public and politicians no longer viewed scientists as infallible; they wanted more results and more accountability. In the U.K. the public was tired of the large burden placed on the taxpayer for the college education of its young citizens, even though enrollments are low – less than 15 percent of 19-year-olds are in universities. However, enrollments at polytechnics, which teach applied science rather than basic research, are increasing, and businesspeople have come to dominate the committee positions once filled by representatives of universities. The shift toward short-term results, patents, and links with commerce, as documented in several chapters here, is moving ahead steadily.

The changes in funding and administering higher education mean a loss in the "climate of free inquiry and cooperative endeavor in which the scientific imagination flourishes," Williams writes. Like Bok, she questions these new arrangements. "Those who want to harness the universities to commercial objectives," she warns, "may destroy the very qualities they admire in them – intellectual excellence, free inquiry, scientific imagination. The pendulum has swung too far."

Maureen Woodhall, a researcher at the University of London's Centre for Higher Education Studies, underscores the points Williams makes with data from an ongoing project examining in depth the funding arrangements of 24 British universities, colleges, and polytechnics. Woodhall's chapter, "Changing Patterns of Finance for Higher Education: Implications for the Education of Scientists and Engineers," also reviews the findings of a study conducted by the Organisation for Economic Cooperation and Development (OECD) in 1988 that identifies university financing trends in 11 OECD countries. Woodhall, a member of the research team, reports that the U.K. and the other OECD countries are moving in the same direction. British industry now contributes seven percent of all university income, with medicine, engineering, and the physical sciences receiving the highest proportion. University management has also changed dramatically to increase control over finances, to generate more income, and to reduce the numbers of tenured staff. And applied research is taking precedence over basic. The OECD study mirrors the findings from the U.K. research. The political climate of the 1980s led to a shift from unconditional public funding toward conditional grants and market demands, Woodhall reports. In several countries the public funding agencies are beginning to see their role as 'buying' academic services on behalf of the community rather than managing or regulating institutions. The OECD study found "an increased proportion of university income derived from contracts with industry and commercial organizations" – a trend several workshop participants noted. The report concludes that basic research will suffer; that commercial investment will lead to problems of intellectual-property rights, which in turn "may restrict academic freedom and may slow the progress of scientific discovery by impeding the full flow of information; and [that] too much academic time [will be] devoted to income generation and bidding for funds at the expense of time available for teaching and research."

Woodhall questions whether competitive market forces – and here she includes competition for fee-paying students – are really an appropriate model for funding higher education. Advocates of market mechanisms believe those changes will enable universities to become more efficient and more responsive to new scientific and technological developments. Woodhall, herself less optimistic, agrees with those critics who worry that the university cannot serve the market and maintain its independence and quality. Indeed, the directions in which S&E university funding and control are moving in the 11 countries in the OECD study are convincing evidence not only that major changes are under way in universities, but also that the repercussion will be deleterious. Williams does admit that this decisive shift in funding and operation has brought some gains – increased entrepreneurial skills for researchers, a greater awareness of market forces, and a heightened sense of what can be commercialized. But she also believes the quality of basic science has suffered and that the future for free inquiry and collaborative efforts is gloomy.

## Current Trends in the Education of S&Es

The aforementioned chapters, each with a different perspective of the changing university, provide the context within which to examine S&E human resources. The workshop participants returned again and again to the subject of S&Es. How many would be needed in the next century? Would the numbers be adequate? Where would they come from? Would their educations allow them the flexibility to move into new fields or employment opportunities?

At a recent international conference on engineering education, participants from Europe, the U.S., and Japan agreed that the combination of a decline in the birthrate coupled with problems of an aging work force will lead to major problems in the supply of S&Es. The international group agreed, "We are probably approaching the most competitive decade in history with a decreasing capability in skilled labour!"[9]

In the U.S. a shortage of engineers has led to an increasing dependence on nonnationals. At present approximately 60 percent of all assistant professors of

---

[9] Tegart, op. cit., p. 25.

engineering under the age of 35 in U.S. higher education institutions are nonnationals. (Nevertheless, almost 9 percent of full-time engineering faculty positions are unfilled.) In 1990, 57 percent of all mathematics Ph.D.s in the U.S. were awarded to overseas students.[10] In the future the number of foreign-born scientists is likely to decline as many of the S&E baccalaureate graduates that came to the U.S. for graduate work are now returning to their home countries. Karl Willenbrock of the National Science Foundation has observed that the decline in immigration has already begun: The percentage of immigrants who are S&Es dropped from 3.6 percent in 1970 to 1.7 percent in 1988.[11]

Even if the birthrate rises, the interest in S&E in the U.S., the U.K, and to varying degrees other industrialized countries is dropping. In the U.S. the college-age population is expected to decline from 30 million in 1982 to 24 million in 1995. Currently only five percent of U.S. citizens earn B.S. degrees, and of those, some 80 percent are white males. Even though the numbers of 18- to 24-year-olds will rebound by the late 1990s, the gains in the population will be from groups that traditionally do not chose careers in engineering, namely women, blacks, and Hispanics.[12]

These concerns are reflected in the chapters in this section on the education of foreign S&Es, the supply problems of the next three decades, and the need for Ph.D.s in Japan.

Dorothy S. Zinberg's chapter, "Contradictions and Complexity: International Comparisons in the Training of Foreign Scientists and Engineers" reports the partial findings from a study of the social, political, and demographic aspects of the education of foreign S&E students, particularly those in the fields of condensed-matter physics, electrical engineering, and biotechnology in France, West Germany, Japan, the U.K., and the U.S. It also explores the dilemmas of countries such as China that are dependent on science education from abroad but fearful that their students will not return.

The number of international foreign students is significant. More than one million (excluding postdoctoral students) travel abroad each year, more than one-third to the U.S. Approximately one-half study science, engineering, and other technical subjects. This activity alone, which has increased by a factor of 10 since 1950, has contributed greatly to the internationalization of science and technology. In addition, the search for scientific talent and ideas, whether in academia or industry, has become global.

The foreign-student study reveals an international tradition beset with problems. In the U.S., where the issue has become politically volatile, the University of California, for example, has agreed to limit the number of foreign students in engineering. In the majority of industrialized countries foreign S&E graduates are enjoined from employment; some countries want to preserve professional jobs for their own citizenry, and some want

---

[10] Keith R. Wilkinson, *S&E Personnel: A National Overview Special Report*, (National Science Foundation, NSF #90-310), p. 29.

[11] With thanks to Karl Willenbrock for making these and other NSF figures available.

[12] Dorothy S. Zinberg, "The Next Generation of Engineers: A New Breed?" *Aspen Institute Quarterly* 2 (Spring 1990): 5,057.

to avoid sender countries' displeasure for stealing their best talent. In all instances, the policies and practices of the leading host countries reveal the tensions between nationalism and globalism, intellectual property and the universality of scientific information, brain drain and brain gain from less developed to industrialized countries.

Alan Fechter's chapter, "Science and Engineering: Human Resource Needs in the Next Three Decades [U.S.]," argues that despite well-documented demographic statistics – a 28 percent decline in the birthrate between 1957 and 1973, the underrepresentation of women and minorities in science and engineering, and the increasing median age of S&Es – results from studies based on population size must be treated with caution. The NSF's and others' predictions of a shortage of 675,000 S&Es by the year 2006 are likely to be overstated, he charges. Fechter's data reveal that United States S&E undergraduate-degree production will indeed decline. Yet he points out that, if encouraged, more bachelor-degree holders could continue on to Ph.D.s, thereby diminishing the shortfall. In addition, most predictive models do not take into account market forces that could correct the situation, potential changes in student-faculty relations, and greater fungibility of the S&E Ph.D. Fechter's data demonstrate the need to develop more sensitive measures to assess the supply-demand imbalance and, as he states, to do so with more humility, because projections are at best fraught with uncertainties.

Fumio Kodama and Chiaki Nishigata of Japan believe the shortages, particularly in engineering, will be substantial. In their chapter, "Structural Changes in the Japanese Supply/Employment Systems of Engineers: Are We Losing or Gaining?" the authors delineate a number of problems, which they believe are germane not only to Japan but also to other industrialized countries. They demonstrate the growing gap between the need for engineers – especially mechanical, electrical/electronic, and materials – and the numbers who are choosing to study engineering. In Japan in 1988 the numbers and percentage of engineering bachelor-degree graduates dropped drastically. Many of the most highly qualified graduates are turning away from manufacturing, once the major employer of engineering graduates, to finance and insurance, the major service industries. Young graduates reject manufacturing engineering because the factories are outside the major cities and the work is dirty. However, the recent downturn in the stock market and the collapse of real estate development are expected to lure young engineers back to traditional employment.

In the authors' comparison of U.S. and Japanese education and employment patterns, they note that official Japanese statistics suggest that the "economic miracle" was accomplished with only a marginal rise in the number of S&E doctorates. This misconception arises because one form of Ph.D. is omitted from the statistics. The dissertation Ph.D. is awarded after candidates complete coursework at a university; they write a thesis only after completing work carried out in an industrial research laboratory. Combining the figures from this group with the traditional university-based engineering Ph.D., Kodama and Nishigata demonstrate that the number of engineering Ph.D.s granted between 1970 and 1986 actually rose 75 percent. The so-called miracle was indeed fueled by a rise in Ph.D.s, not by master's degree production alone as frequently reported. However, the steep increase has ceased. In 1965 some 33 percent of master's degree holders went on to obtain Ph.D.s of both types; by 1989 only 9 percent matriculated, and 30 percent of those dropped out before graduating. The master's degree remains the most common level of education in engineering. Kodama and Nishigata believe dissertation Ph.D.s could be an important factor in Japan's economic vitality. The rigidity of Japanese university departments makes it difficult to forge the new, interdisciplinary groups needed to spearhead innovation in the information technologies. Industry, on the other hand, operates under no such constraints and hence allows students to work at the cutting edge of new fields.

Throughout the workshop discussions, the participants overall agreed that industrialized countries will need increasing numbers of S&Es in the decades ahead, more than could be expected from the current rate of enrollment of undergraduates in S&E. The disagreements focused on just how many might be necessary to keep science and engineering a vital enterprise.

Greater consensus occurred on the future of the technical work force, that is, the non-college-educated but technically trained work force. The U.S. Bureau of Labor Statistics predicts that between 1986 and 2000 there will be a 76 percent increase in computer-related jobs. Yet as Harvey Brooks's chapter, "Educating and Training the U.S. Work Force for the Twenty-First Century," demonstrates, there is a growing mismatch between the skills of the labor force and employment opportunities.

Unlike college-educated S&Es, particularly those trained in research universities, where the problem is quantity rather than quality, the technical work force in the U.S. lacks "the basic language and math skills necessary to learn the new skills required by technology and market structure." Learning on the job is becoming more difficult as "new generations of production technology [will] succeed each other so rapidly that specialized skills learned primarily on the job will become obsolete several times within the working life span of the typical worker." Many industrialized countries, for example, Germany, Sweden, and Switzerland, have significantly superior job-training programs for their populations. But even they are beginning to worry about the adequacy of these programs as each nation becomes culturally more heterogeneous and faces increasing pressure to absorb a massive wave of immigrants fleeing the devastated economies in eastern and central Europe. Brooks advocates not only increased scientific literacy for the work force, but also university contributions to the larger understanding of educational strategies and "a much deeper theoretical understanding of the learning process."

## The Flow of Information

The product of the university is knowledge, which to be effective must be communicated. In this section two authors discuss the need to strengthen the international mobility of S&Es in order to enhance collaborative research efforts; and the impediments that government, industry, or even individuals can impose on the communication of information that might provide a competitive edge to other countries. Both authors state that new electronic communication technologies make such tactics self-defeating. New policies are needed to deal with the fact that information cannot be hoarded. The question becomes, "How can it be shared?"

In "The Changing Patterns of International Collaboration in Universities," Jean-François Miquel of the Centre National de Recherche Scientifique in Paris examines the increasing importance assumed by international coauthorship and cooperative efforts in the generation of scientific knowledge. Miquel argues that despite advancements in information technology that facilitate instant communication and data sharing, "personal contacts and visits abroad are becoming more and more indispensable." Academics worldwide have long recognized the importance of this finding. Scholars have traveled to foreign universities and centers of learning as long as there have been written records. Symbolically, a successful European program begun in the 1980s, the European Community Action Scheme for Mobility of University Students, took its acronym – ERASMUS – from a fifteenth-century Dutch scholar who journeyed to Siena, Italy; Cambridge, England; and Zurich, Switzerland to carry out his work with other scholars. When the program ERASMUS was initiated, the founders noted that in the twentieth century proportionately fewer European students were studying in other EC countries than in Erasmus's time! Today some 50,000 EC students are enrolled in other member states' universities, and the numbers are rising rapidly.

Beginning as foreign students in university laboratories or as postdoctoral fellows, young scientists later forge collaborations with former professors and fellow students. The exchange of postdoctoral students is also a major pathway to international collaboration. Miquel's research yields several analytical insights with direct relevance to policy. One is methodological, the Science Citation Index technique, which quantifies the internationalization of scientific activities by charting the frequency of references to coauthored and collaborative research publications in the scientific literature. In addition, Miquel presents two empirical, macro-level case studies of EC countries, Denmark and Greece, and demonstrates their affinity toward the U.S.; and a micro-level analysis, which focuses on individual cooperation among scientists from person-to-person exchanges and collaboration. The policy implications are significant. One of the victims of decreased funding for science has been foreign-research-related travel, workshops, and international conferences. The recent surge in worldwide oil prices and its effect on airfares can only exacerbate the significant barriers already in place. The research reported here suggests strongly that to keep science, engineering, and technology vital, international scientific cooperation has to be supported not only through the exchange of articles and telecommunications but also through person-to-person contact.

In the chapter "National Security Information Controls in the United States: Implications for International Academic Science and Technology," John Shattuck reviews the history of national-security information controls of the past two decades. He argues that restrictive U.S. policies may be counterproductive and expresses hope that the recent changes in Eastern Europe and the Soviet Union will encourage the government to loosen them. However, he writes, concerns about economic competition with the EC could spur policy makers to tighten restrictions where intellectual property developed in universities is involved. Several impending changes can begin to improve the climate for scientific cooperation – less restrictive national-security export controls; changes in the immigration laws to increase the flow of foreign S&Es to the U.S., especially from Eastern Europe; and fewer restrictions on the export of high technology and scientific information. Shattuck cites seven incidents that took place in 1980, when the U.S. Departments of Defense and Commerce attempted to limit attendance at professional scientific meetings to U.S. and Canadian citizens and permanent U.S. residents. Some scientific and technical societies themselves have informally barred foreign scientists from their meetings – a practice prohibited by some two-thirds of professional scientific societies surveyed by the AAAS. Shattuck agrees with the findings from four National Academy of Science studies that warned of the harmful effects of broad controls on U.S. economic and military security.[13] "With respect to U.S. military and economic progress," the 1982 report concludes, "controls may slow the rate of scientific advance and thus reduce the rate of technological innovation. Controls also may impose economic costs for U.S. high-technology companies that offset both their prices and their market share in international commerce. Controls may also limit university research and teaching in areas of technology. A national policy of security by accomplishment has much to recommend it over a policy of security by secrecy."

Shattuck proposes that the federal government reduce caps on the number of nonimmigrant scientists allowed into the country and speed up labor certification for academic employment of nonnationals in order to strengthen U.S. universities. In addition, universities should address the issues of intellectual property and information control by "adopting flexible policies that deal realistically with the entire range of potential conflicts," he says. Like Bloch, Bok, and Armstrong, he believes the increasing value of intellectual property generated by universities is bringing about changes that have to be planned for in order not to distort their mission.

## Summary

In summary, the demands being placed on the university both from outside and from within have forced it to change in new, not fully understood ways. These conflicting demands include:

– the need for openness in scientific research and the tightening of laws concerning intellectual property;

– university efforts to internationalize the student body and the proprietary demands of industry and government;

– traditional peer review and the direct appeal to Congressional representatives for funds in the U.S.;

---

[13] "Scientific Communication and National Security," 1982; "Impact of National Security Controls on International Technology Transfer, 1987; "Global Trends in Computer Technology and Their Impact on Export Controls," 1988.

– the push toward global scientific collaboration and the struggle for national preeminence; and

– the identification of the S&E work force as a national resource as well as a global commodity.

The workshop papers demonstrate the tensions between these conflicting demands as well as their interdependence. The participants called for new networks of international representatives from each of the institutions involved. The next step must be the formation of proposals for global studies and policies. National policies, even joint efforts, are no longer sufficient. The future of the next generation of S&Es can be better assured by a multinational examination of values and attitudes, demography, and the global mobility of talent. And the future of the university can best be assured by a cross-national study of the complex relationships among industry, government, and universities worldwide. Policy for human resources and universities has to become more complex in order to deal with a new global system. National identity alone will no longer serve a country well.

The workshop ended on a note of optimism for the possibility of resolving some of the contradictions that had marked the discussions. The next meeting will aim to draw science-policy practitioners and professionals with related interests from more countries. The proposal is to meet next in the center of the new Europe, Berlin.

# RECONCILING CONFLICTS – THE CHALLENGE FOR THE UNIVERSITY

DEREK BOK
*President, Harvard University*
*Massachusetts Hall*
*Cambridge, MA 02138*

As societies become more advanced and more complicated, universities and the education and research they provide grow steadily more important. Certainly, we at Harvard – as at other universities – experience every day our increasing involvements with industry, government, and other important segments of society. While these links bring many wonderful opportunities, they also create many complicating entanglements and difficult dilemmas that have provoked much debate.

From my limited vantage point I can only say that much of the discussion in the United States on these topics leaves me dissatisfied. One of the things that concerns me is the persistent tendency to emphasize too heavily, in my view, the importance of relationships between universities and industry as a means of getting ahead competitively. The validity of this linkage is far from clear. If universities are really the key to greater competitive problems, why did we grow faster than other countries during the late nineteenth and early twentieth centuries when our universities were mediocre, and why is our productivity lagging behind other nations now that our universities and their scientific achievements lead the world? I think the answer is that university research and its linkage to industry are not decisive factors in the competitive problems of the United States. I fear that if we continue to try to justify our research in those terms, to try to exploit competitiveness as a reason for building science and links with industry, we are likely to end by disappointing our constituents to our ultimate disadvantage. I also fear that we may unwittingly and unwisely favor restrictive policies toward foreign students and foreign companies in a misguided effort to gain a competitive advantage. In particular, I worry that we will encourage exclusionary policies and secrecy requirements by seeming to suggest that what universities do is really the key to U.S. competitive resurgence.

Rather than pursue this misguided course, I hope that we in universities always remember that the great utilitarian purpose – if we need a purpose for science – is the welfare and betterment of all humankind and not the betterment of any particular society at the expense of others. If we stray from that path, I believe that we will eventually be in difficulty.



*D. S. Zinberg (ed.), The Changing University*, 15–16.

The other thing that concerns me is the tendency of different groups to talk past one another about the proper relationships of universities to society, government, and industry. I see one group of people, particularly in government, speaking about the need to create closer links with industry to speed the application of knowledge to useful ends, while other critics attack universities for compromising their values by cozying up to corporations. I see much effort to bring in bright foreign students, many of whom stay and help us achieve an adequate supply of scientists and engineers, while other groups criticize universities for developing relationships with companies overseas, which can be viewed in large part as another way to educate foreign scientists by allowing them to come and see what we are doing in our laboratories. I see prominent people in government, business, and foundations constantly trying to lure our professors to spend time giving them advice, while the same people in the guise of parents, criticize faculty members for not spending more time on campus teaching their children.

The point is not that these conflicts are unnatural or wrong. But we should be spending a lot of time carefully sorting out our often contradictory expectations and demands, trying to reconcile them in some internally consistent way, so that we can develop reasonable standards by which to judge the performance of our universities and hold them accountable.

At present, careful discussions of this kind are all too rare. Instead, I hear people on both sides of these debates speaking to only one side of the issue and perpetuating rather than reconciling the inconsistencies. I can understand this tendency in the media, because you don't build circulation or increase television audiences by delicately balancing conflicts and reconciling inconsistencies. But university administrators, faculty members, and high public officials ought to do better than that. Those who care about universities should devote more time to the process of reconciling these contradictions. In doing so, they should be simultaneously aware of the social need and obligation for closer connections between universities, governments, and corporations, as well as the compromises, the dangers, the entanglements, the long-term risks to the vitality of basic science if appropriate safeguards are not put in place.

# AN INDUSTRY PERSPECTIVE ON THE CHANGING UNIVERSITY

JOHN A. ARMSTRONG
*Vice President for Science and Technology, IBM*
*Armonk, NY 10504 USA*

## Abstract

The globalization of high technology business is changing the industry/university interaction. Five important consequences are identified. First, the responsibility of U.S. industry for university support and the role of foreign-based corporations are not clear and require thoughtful discussion by those in government, industry, and academia. Second, universities should assess whether their role in technology transfer and national competitiveness has been correctly stated. Third, the university view of intellectual property rights and protection needs alteration, in many cases, so as to avoid impeding technology transfer and cooperative research. Fourth, many universities insist on a double standard of conduct when economic matters are at stake. Conflicts of interest arise as faculty accept public funding to invent and then involve themselves in enterprises that exploit these inventions in the marketplace. Fifth, given the unprecedented reliance of U.S. industry on foreign-born scientific and engineering talent, consideration should be given to enhancing the university graduate curricula to help prepare foreign students for managerial leadership.

## Introduction

From the end of World War II to the late '70s, the U.S. economy was dominant in the world. The U.S. had the best-educated work force and the highest productivity. It led the world into the age of information and microelectronics. It helped Europe and Japan rebuild their economies. Most U.S. businesses, even those that were international in scope, defined success in terms of the home market.

The '80s saw major changes that affected both business and the universities. U.S. dominance in manufacturing and technology development slipped away in important areas. The finance and capital markets, much of the world's manufacturing and almost all of high technology business have become globalized. Success for many enterprises has become dependent on successful operation in all the world's major markets. The logic is simple: If one does not meet one's competitors in, say, the European market, competitors will use profits in that market to finance their attacks on you in the market where you are strong. These developments have had profound effects on the strategies of corporations.

In the past, when either industry or academia has changed, their relationship has



*D. S. Zinberg (ed.), The Changing University*, 17–23.

changed. Much more profound, perhaps, will be the change in university/industry interaction when both are in the midst of change.

In this paper I focus on what has changed, or will have to change, in the university in response to the globalization of economies. There is clearly a movement to expand the idea and the reality of the global university. However, although the ethos, the intellectual reach and the recruiting of faculty have long been internationalized, the financial base of the university is not international. On the contrary, it is often aggressively national. The achievement of a more truly international function for our universities will be difficult. As long as university funding is overwhelmingly national, their progress toward that function will be severely limited. Many industrial companies are more nearly globalized than any university in financial support, labor and facilities.

As a scientist, I believe in the important role of our great research universities; as an industrial executive, I am convinced of the importance of mutually beneficial interaction between the universities and industry. And, as a friend, I want to raise five issues concerning the current situation of U.S. research universities.

I raise these issues fully knowing that they arise out of day-to-day work in the world of industry, and that they do not represent either scholarly investigation or a broad view of industry as a whole.

Here are the five issues in brief:

– What can and should be the responsibility of U.S. industry for support of the universities? What will be the universities' response to offers of support from corporations not based in the U.S.?

– What is the appropriate role of the university in technology transfer and national competitiveness?

– What is happening in the universities concerning intellectual property rights and protection?

– Do some university administration and faculties follow a double standard of conduct when matters of economic value are at stake?

– Is the present graduate curriculum preparing foreign students not only to make technical contributions, but also to share in the managerial leadership of U.S. industry?

## Financial Support by Industry

New financial needs are forcing a reexamination of the responsibility of U.S. industry for support of the universities. They have also brought into focus the question of what support corporations not obviously American should give.

Since World War II, U.S. universities, primarily with the support of the federal government, have pursued frontier research in science and engineering. But there are enormous pressures to reduce both federal and state expenditures. The causes for these pressures are many: an expanding array of social ills, the decline in real income growth (and thus tax revenues) due to lagging productivity, etc. Whatever the causes, the pressures are real, the national debt is increasing, the federal deficit is not being brought under control. As a result, funding for education and research is under extreme pressure. Although overall federal funding for university research has recently been increasing strongly, the funding per investigator has not kept up with the opportunities or with inflation.

Industry/university relationships have changed remarkably over the past decade in the U.S. and, to some extent, in Europe. One result of the new closer relationship has been the belief among some in academia that industry will step in, or should step in, to make up the shortfall in federal support for research. This is unrealistic.

U.S. industry, too, is under increasing pressure to sustain earnings in a fiercely competitive global market. At the same time, rapidly advancing technology requires equally rapidly growing levels of investment. In the electronics industry a generation of memory chips has a three-year lifetime. The current generation requires an investment of between $100 and $300 million to set up a manufacturing plant. By the end of the decade the cost will be about $1 billion. Total corporate spending on research and development, which was increasing by a meager 2 percent in constant dollars in the past few years, actually went down by 1 percent in 1989.

Outside the U.S., industry's main financial contribution to university funding is via taxes paid. But also in the U.S., industry by and large believes that it contributes its main share to the support of universities through taxes and through existing corporate philanthropy. Future additions to industry-funded programs are likely to have a relatively narrow focus. It is not realistic to expect that further strengthening of industry/university relations will be built primarily on greater financial support; at least, this is true for those industrial sectors that are under great competitive pressure.

Support for U.S. universities by non-U.S. based corporations is growing. Regardless of the fact that overseas corporations have supported U.S. universities for years, it has now become a very divisive issues. This is because both in the federal government and in industry, there are deep differences of opinion among decision makers concerning whether foreign support of research in U.S. universities is in the national or the U.S. corporate interest. It is likely that universities will have to choose, in some cases, between new support from foreign sources and continued support from existing U.S. sources.

In raising this issue, I am aware that I cannot give the right answer. But it would be a mistake to ignore the substantial ignorance and lack of sympathy, both in corporations and in government, for the ethos of the open, internationally-minded university community. Much thoughtful discussion, both listening and explaining, is necessary before policy consensus is possible here.

I do not believe that we have yet found the right forum for bringing the key players together to have the required discussions. This may be one of those areas where we have to solve the problem in practice rather than in principle.

Ironically, this problem is worse than it needs to be, due to a successful campaign by some university people. Over the past few years, they promoted the view that strengthening the universities was an important and relatively short-term way to improve U.S. technological competitiveness.

## Universities and Competitiveness

In my view, the current competitiveness problem in parts of U.S. industry are due to a lack of attention to quality by both management and by workers, lack of proper investment strategies, high interest rates caused by structural economic factors that make it difficult to get the return on investment expected by U.S. financial markets, poorly educated workers, insufficient attention to quality of recruiting, and many other factors as well.

Poor technology transfer from the university or national labs to industry is not a significant cause of the competitiveness problem. And there are only a few sectors, for example, software development and perhaps biotechnology, where technology transfer can have a truly short-term effect.

The often-mentioned failure of U.S. industry to capitalize on its own progress in basic research is real, but it cannot be cured by more research or, to any great degree, by any activity that takes place primarily in the university community.

The university research infrastructure needs strengthening for good and sufficient long term reasons, but doing so in the hope of helping competitiveness in the short term is like trying to cure constipation with force-feeding.

There is, however, one aspect of this problem where the universities could be of major help in the long run: that is in helping Western culture to get rid of, once and for all, the intellectual hierarchy in which "pure" is somehow better than "applied," in which physics is better than chemistry, both are better than engineering, and the discipline and intellectual content of manufacturing is hardly valued at all.

The most compelling reason for discarding this dated intellectual hierarchy is that it does not correspond to reality. Nature knows nothing of the distinctions we make in the university between disciplines or between the various approaches to technical knowledge.

## Intellectual Property

At a time when all concerned parties want to facilitate technology transfer from the academic realm to industry, a new roadblock is approaching.

Questions about intellectual property policies in academia are often raised in the context of the unwelcome restraint on rapid publication, which sometimes accompanies industry-sponsored work with proprietary overtones. But, in my experience, this issue of withholding publication until a patent application is filed can be dealt with by prior agreement on the amount of time in which an application must be filed, and by timely attention to this schedule by faculty inventors and the concerned patent attorneys.

Two other important considerations – not related to secrecy – are often overlooked by U.S. universities in their patent policies.

First, academic patent policies do not take into account the fact that different industries have evolved very different strategies and practices in their own use of intellectual property. Consider the differences in practice between the electronic and computer industries, on the one hand, and the chemical and pharmaceutical industries on the other hand. In computers and electronics, the primary use of patents is to ensure freedom of action. Most large companies in the world are extensively cross-licensed with each other. Exclusive licenses are almost nonexistent. The key is not ownership, it is access.

In the chemical industry, however, a patent on a new material is used to ensure that the patenting company can have the exclusive right to bring the new material to market.

Consider the implications for industry/university collaborations when universities model their patent practices on those of the chemical or pharmaceutical industries. When IBM research or development labs undertake joint studies with university colleagues to the scientific benefit of both parties, we are met time and again with university policies that insist on the university having ownership of any jointly produced patents, moreover, with the aggravating claim that we should pay royalties to use work we have jointly funded and produced. We insist on royalty-free, nonexclusive licenses to work we have cosponsored; major universities persist in applying policies more suited to the chemical or bioengineering worlds.

Our approaches to patent protection differ in other important ways. In large companies, individual inventors usually have no first-order stake in the financial consequences of their inventions. They assign their rights to the company. Professors, in my experience, when they think of patents at all, think of them in terms of the possibility of getting rich, much the way folks play the lottery, and with as much statistical hope of success. This unrealistic expectation of financial interest sometimes leads to unrealistic behavior by university scientists, at least in fields such as electronics and computing.

In those fields, companies want freedom of action. Universities, whether faculty or administration, have really only one interest, and that is to get as much money as possible from licensing patents.

These differences constitute a real impediment to cooperation in important cases and need to be discussed. I propose a university patent policy that is flexible enough to deal with the policies and practices that are normal in the industrial sectors interested in a particular invention.

## A Double Standard?

Do universities insist on a double standard of conduct when matters of economic value are at stake, allowing and even promoting conduct on the part of faculty that would not be tolerated in industrial or federal laboratories?

Professors of electrical engineering or computer science commonly get their research funded by the federal government and, based on that research, some of them form companies in which they may have substantial personal financial interest, either as corporate officers or as major stockholders. They then proceed to compete seriously with other companies that do not have their researchers supported by grants for basic and applied research. These practices go well beyond the standard one day per week that universities allow faculty members to use for consulting and other extramural activities.

There is a paradox here. Some faculty members want the security of tenure, the privilege of academic freedom for pursuing their own ideas, public or industrial money to support their research, *and* full rights of ownership and consequent economic benefit. Clearly, direct conflicts of interest arise as faculty take advantage of public funding to invent and then set up their own companies or become partners in ventures to exploit these inventions in the marketplace.

I am aware that this issue is treated very differently in different universities, but some universities are going to give universities as a whole a bad name and call down unfortunate consequences on all. I personally am outraged that as an individual and as a member of my company, through taxes paid in both capacities, I subsidize university colleagues to compete with me.

This issue needs more discussion, including more discussion in Washington.

## Training Students from Other Cultures

Another major element in university/industry relations is the changing demographics of students in science and engineering. Only a small fraction of our U.S. high school graduates pursue a college degree in science and engineering. In contrast, in other countries – especially the Far East and developing countries – math, science and engineering are cultivated as most desirable for the best and the brightest. As a result, the fraction of foreign students in science and engineering graduate programs in U.S. universities has been increasing rapidly. The newest trend is that even undergraduate programs in science and engineering are attracting an increasing number of foreign students.

It is widely viewed as a weakness that the U.S. has a growing dependency on imported scientists and engineers. However, the reality is that for the foreseeable future U.S. high technology does have such a dependence. (Similar situations may arise in other countries, too.) If only a small number of graduating foreign students return to their country of origin, the U.S. will continue to prosper. On the other hand, large scale repatriation would be very detrimental because of a double loss: the investment in the education of foreign students (usually U.S. government or industry funds for graduate education and research) is lost to our economy, and U.S.-educated scientists and engineers working in other countries become a competitive weapon against the U.S. These worries are likely to be heightened more as Congress seeks a bigger role in determining how and which people are supported by federal research monies.

Universities should join forces with industry in lobbying for automatic work permits for any foreign student with an advanced degree from an accredited university who wants to stay in the U.S. to work. This is clearly a remark made in the U.S. context; other countries will see the matter differently, but competition for talent is one of the consequences of globalization.

Given the unprecedented reliance of U.S. industry on foreign-born scientific and engineering talent – a reliance that is quantitatively greater than ever before and sure to be of longer duration – can U.S. graduate school curricula prepare these students both to make strong technical contributions and to lead as managers in U.S. industry?

Foreign-born scientists and engineers sometimes face cultural difficulties when they try to qualify for management positions in the U.S. In many cases technical skills may be less important than communication skills based on informal networks, the shared experiences of youth – in short, the skills based on the shared culture dominant in this country. This is a subtle problem, but, because of the magnitude and the likely long duration of U.S. reliance on foreign-born talent, it raises important questions for university policies. Should U.S. universities provide only technical education, or should they also help to prepare foreign students with the communication skills and cultural background needed to succeed in the U.S. business world? Is this an area where industry and academia can work together?

In summary, the university, like industry, faces many challenges. In some ways, corporations are more advanced than the universities in evolving to meet these challenges; however, both industry and academia are better prepared than the U.S. federal and state governments to understand and respond. Overseas, the situation is different. Governments sometimes take the lead. In any case, it will be important to continue this discussion in diverse settings and contexts.

# POSITIONING U.S. SCIENCE POLICY FOR THE NEW WORLD ORDER

ERICH BLOCH[1]
*Former Director,*
*National Science Foundation*

## Abstract

As world economic competition increases, countries must cultivate their science research programs, the communication of results, and the education of their work forces. Unfortunately, the same forces of economic competition can cause governments to institute policies that endanger these priorities. Protectionism may limit international access to research results. An insistence on tangible, profitable results may endanger basic research programs. Because science depends upon the abilities of scientists, the U.S. must find ways to make careers in science and engineering more attractive to its citizens.

## Introduction

The Chinese symbols for "change" are "danger" and "opportunity." They reveal the wisdom of an ancient culture. And because we live in times of monumental change, we must confront both the opportunities and the dangers.

Once access to natural resources was the key to economic success; today access to technology is more important. Recent events in Eastern Europe and the Soviet Union result, in part, from the power and reach of technology. As the Cold War fades, economic competitiveness will move center stage. The arrangement of world affairs around a bipolar military contest is being replaced by a global economic competition in which the United States, a united Europe, and the Pacific Rim nations are the major players and the major competitors.

To compete in today's markets requires continual innovation and improvement. To compete as a nation requires

– strong basic research;

– a way to transmit and share new knowledge, that is, cooperation, both domestic and international; and

– well-educated scientists and engineers, and technical work force.

---

[1] Erich Bloch retired as Director of NSF as of August 30, 1990. His current address is 2801 New Mexico Ave., N.W., Washington, D.C. 20007 USA.



*D. S. Zinberg (ed.), The Changing University*, 25–30.

In other words, knowledge and people are the primary economic resources today. They will determine the U.S. position in the new world order, just as they will determine the position of every developed and developing nation.

## University Research and an Economic Resource

In the U.S. we've been fortunate. During the 1960s academic institutions, notably universities, became the single most important researcher in basic science and engineering, and that arrangement proved very successful in producing useful breakthroughs. So far, more than 50 U.S. scientists have received Nobel Prizes for research conducted at U.S. universities during the late '50s and the '60s alone. In the information sciences, university researchers provided the concepts and tested new developments that later became the standards for industry – developments such as timesharing, dynamic relocation, graphics, and visualization. Equally important results in other disciplines have been achieved.

What is particularly relevant to this discussion is the fact that so many academic research discoveries of the postwar decades have had such profound economic effects not only in the U.S., but worldwide. Many common products and services today were the unanticipated results of university research. Whole industries – semiconductors, biotechnology, computers, many material areas – are based on research begun in universities in the '40s and shortly thereafter.

The economic benefits of academic research haven't been limited to the findings themselves. Many of the tools created to advance research – electron microscopes, magnetic resonance equipment, recombinant DNA techniques – have also found profitable application in industrial and commercial sectors.

Fundamental research since the '40s continues to be a resounding economic success story. But in some ways it has become the victim of its own success. Investment in basic research is now made with an expectation of economic return. Research is now considered to be of strategic national importance. Every region and country wants to build academic centers to nucleate economic growth. But by doing so, they may diffuse talent. We must be wary. Moreover, as the economic value of basic research rises, some would like to restrict the flow of scientific information to prevent its exploitation by foreign competitors.

We can understand these concerns. Who expected, in the late '30s and early '40s, that the new insights into the behavior of conduction in solids would be the basis for the transistor and all the microelectronic technology it has spawned? The returns to basic research are often unexpected; they're often realized only in the long term. Often they're difficult to valuate, even retrospectively.

What is not generally understood, except by the practitioners themselves, is that basic research thrives on openness, communication, and cooperation. As a supporter of basic research, the National Science Foundation (NSF) wants to foster such an environment.

Another important consideration is setting priorities among worthwhile alternatives, especially when funding is limited. Drawing on the language of economics, a balanced portfolio is a better approach to maximizing national investments in academic research. We never know what fields will be important, from where or from whom the next new insight will come. We want to make sure that research support is balanced so that no one group, no one field, no one approach consumes too much of the budget. Balance does not mean that all fields of research have an equal claim on public funds, however. It is entirely reasonable to emphasize those fields that offer greater promise of payoff, or are on the verge of rapid and important development, or need the extra push to succeed. Materials science, solid state physics, biotechnology, engineering, and atmospheric chemistry come to mind, because much of the research in these fields is closely related to current social concerns – economic competitiveness, public health, environmental protection – and is equally poised for new and important insight.

In choosing among broad fields or competing projects, government officials who allocate research funds should consider the likely effect on economic or other societal concerns. Nonetheless, it is important to remember that there is no way to know when an unexpected insight in a field unrelated to current public concerns will have far-reaching impact. More than ever before, knowledge in science and engineering is a seamless whole. It is a mistake to consider any field irrelevant.

An overarching task of the government is to ensure a viable infrastructure – that is, the laboratories, equipment, and most important, the scientists, engineers, and students. There is a critical mass of material and human resources below which research will languish.

## Cooperation and Knowledge Transfer

The essence of basic research is the search for new knowledge, discoveries, and inventions. Yet transmitting new knowledge can be as valuable as knowing the results. Broad distribution and discussion of research results increases a nation's capacity to absorb and use new knowledge. Thus, knowledge transfer is an important component of basic research. The time between research result and useful application is getting shorter. Timely and effective communication of new knowledge to those who can use it requires cooperation.

The separation between industry and academic research should be avoided at all cost. Many of the programs that the NSF has initiated in the last five years seek to remedy that shortcoming. Everyone benefits from the cooperation. Innovation is an interactive, iterative process. Rarely does a fundamental research result find a straightforward application. The flow of ideas and information back and forth between originators and users is what creates new products and technologies.

In another vein, society's problems will always be an important source of inspiration for scientific research. Cancer, heart disease, AIDS, earthquake prediction and hazard mitigation, waste management – these problems need immediate clinical and technical solutions.

What is true at the national level is also true at the international level. We ought not let national borders limit our efforts to expand cooperation. For one thing, the cost and complexity of modern research is such that no one single country has the resources to do it all. Sharing at many levels – facilities, instruments, people – must become a regular occurrence in more than the traditional areas, such as astronomy and nuclear physics; but also in materials science, engineering, chemistry, and many more.

But the main reason we need to promote international research cooperation is that many of our pressing national problems are global problems, such as global warming and nuclear waste management. Their solutions require greater cooperation. It's now possible. Day-to-day communication between researchers across the globe is a recent phenomenon made possible by electronic networks, computers, and workstations. We're only beginning to see its effect, but the potential is vast. Rapid diffusion of new knowledge can enrich all nations that promote this new mode of communication. But it will only work if those who participate contribute their fair share and permit a true quid pro quo.

Access to the newest research is a source of competitive advantage, especially when the prospects of early commercialization are real. This situation creates political pressure to limit access to research information. We must recognize these pressures without succumbing to them.

In the U.S., we question whether the single-market momentum in Europe will lead the European research community to disrupt collaborations that our researchers have established over the years. In the past scientific exchange between the U.S. and some countries has in certain cases been greater than exchange between neighbors. Whether the change in economic focus affects research focus is important.

Currently, research and development spending by the European Economic Community (EEC) is less than four percent of total European research and development spending, and EEC cooperation is mainly in strategic technologies with clear commercial potential. On the other hand, cooperation between basic researchers occurs mostly through research facilities sponsored by national governments and through universities. These relationships are the focus of our concern. We're concerned because budgets for basic research in most EEC countries have been declining or are stable. Increases have been directed mainly at technology development. To the extent that this imbalance persists, contacts between U.S. and European researchers might diminish, or there might be pressure to align international research cooperation with technology development activities.

Will momentum of research in strategic technology by the EEC begin to affect the allocation of resources and with it access of U.S. research to national programs in Europe? Will intra-European connections lessen the commitment of individual nations to the U.S. research community over time? What role will the EEC play in bilateral research decisions? These are the questions U.S. science policy makers must ask.

No nation has a monopoly on knowledge. Now more than ever, research institutions, both in academia and industry, must reach out and draw ideas and inspiration from foreign sources.

In the U.S., we need to work harder to strengthen and expand our foreign ties. Compared with other researchers around the world – especially those in Europe and the

Pacific Rim – Americans have been slow to appreciate the value of foreign endeavors. We have been slow to seek research opportunities abroad, relying more on others to establish and promote collaborations. We can no longer afford this approach. Universities might want to restore the tradition of foreign research that was part of U.S. graduate education earlier in the century. There are many opportunities. U.S. researchers have much to contribute, but they also have something to learn.

## The Importance of People

New knowledge is necessary. And how we cooperate to create, transmit, and use new knowledge is important. But the crucial part of a research endeavor are the researchers.

People do research – not machines, instruments or facilities. And the quality of the people drawn into the basic research enterprise is the best predictor of its success. If there is one key lesson to be learned from the success of U.S. academic research in the postwar decade, it is that new people, new perspectives, and new ideas make research vigorous and vital.

The influx of European talent during and after the war and the remarkable growth in the number of new scientists and engineers educated in the '50s and '60s under the G.I. Bill have fueled research creativity in the U.S. over the past decades. By coupling research and education, U.S. universities have attracted scientific talent from around the world, strengthening our own and worldwide research capabilities.

The merit of this arrangement – tying research to education, making the university the locus of the nation's basic research effort – cannot be overestimated. The complex task of basic research requires more than knowledge. It requires immersion in each unique laboratory culture. Students hone their understanding and absorb the culture as they work with seasoned researchers. Thus, each new generation of scientists and engineers learns what constitutes good work; they learn what's expected. That sets the standards for research and technical performance in all sectors of the economy.

In this context, the number and quality of a nation's students demonstrate the vitality of its basic research enterprise. In the U.S., the NSF has noticed some disturbing signs. Young people are turning away from careers in science and engineering. For every graduate physical scientist or engineer joining the work force today, there are two MBAs; at the undergraduate level, four business majors graduate for every three graduates choosing a physical science or engineering major. Twenty years earlier, the ratios were reversed.

We must ask ourselves, why are today's students opting out of scientific research and technical careers? The reasons are many; some refer to a lack of preparation, others to a lack of economic incentives. For example, starting salaries for research scientists and engineers have not kept up with those in other professions, such as law, business, and medicine. With the cost of a college education rising faster than the cost of living, it's not unreasonable for students to choose fields that promise a better return on their investments.

Moreover, the apprentice period for research scientists and engineers has lengthened. The average time lapse between bachelor's and doctorate degrees is now about eight years, one year longer than a decade ago. And postdoctoral studies are becoming routine in many research fields. Since students forgo earnings during this entire period, is it surprising that they are foregoing further studies?

Maybe we can shorten the time it takes to prepare students for research careers and thereby ease their financial burden, without impairing the quality of the education? We should look at the total educational system to find out what can be done about this issue.

Learning is a continuum. It is a mistake to focus on only one stage in the process. Supporting the research infrastructure means supporting math, science, and engineering education along the entire pipeline. Continuity of support is also important. Employment and career opportunities in basic research crucially depend on the level of public support. Uncertainties about research and development funding can discourage students from pursuing advanced studies in science and engineering.

The plan of Presidents Reagan and Bush to double the NSF budget over the next five years is a positive step toward easing some uncertainties about research careers in the U.S. Steady public support of academic basic research does more than simply fund projects; it creates an environment that attracts talented students to research careers.

## Conclusion

People, cooperation, basic research – these are the strategic resources in today's global economy. We no longer live in a world where the ability to exploit scarce natural resources assures economic success.

People, cooperation, and basic research are renewable, basically limitless, multiuse resources, available to all. So the competition has heated up. And the rules of the game are changing. New competitors – not just new nations, but new national power blocks, multinational companies and transnational industrial consortia – are entering the market. And old competitors are realigning. This also means we need to rethink our science and technology policies and strategies for optimum results.

# THE UNIVERSITY – AND PARTICULARLY THE TECHNOLOGICAL UNIVERSITY: PRAGMATISM AND BEYOND

GEORGE BUGLIARELLO
*President, Polytechnic University*
*333 Jay St., Brooklyn, NY 11201 USA*

## Abstract

The university is an institution that has survived by avoiding too sharp a definition of what it is and what it does. This has led to considerable diversity among universities, and at the same time to a common inability, intrinsic in the design of the university, to respond effectively to major social problems. One special kind of university, far more focused and capable of responding, is the technological university. A case in point is how Polytechnic University is developing a close interaction with industry and a major urban university-industry park in New York City focused on the service industry.

## Survival, Social Role and Ideology

The university is and does many things but, like most other organizations, it wants first of all to survive. The university has become adept at survival, to the point that it is not always clear whether the great range of activities in which many universities are engaged today represents a deep ideological commitment or simply a manifestation of the need to survive. There is obviously much more to the university than just survival, so that the university, in truth, is a mixture of idealism and pragmatism. Some universities, such as Harvard in this country, were created originally to educate the clergy. Other universities, such as Bologna and Padua, were created for the ideal of learning – pure and simple – and were fountainheads of secular learning. The land-grant colleges, were developed not so much because of the importance of learning, but as economic tools. Other universities yet, often again land-grant colleges, had to respond to another need – providing their students with skills, even if those skills are not directly relevant to the economic development of a state or region. For example, when I was at the University of Illinois, one of the debates was whether the university should have an aeronautical engineering department, since there was no aeronautical industry in Illinois. It was concluded that there should be such a department, because the purpose of the university was to give opportunities to the citizens of the state to learn subjects of interest to them, to empower them to pursue careers anywhere, not only in the state.

The university is also idiosyncratic, first of all in the way it operates. It is a miracle that the university can respond – let alone respond rapidly – to anything, given the complex constituencies that somehow must be channeled to reach a



*D. S. Zinberg (ed.), The Changing University*, 31–37.

common agreement. The university is also idiosyncratic in what it considers important. This varies from university to university and from time to time. For instance, for a long time engineering was not considered belonging in the university, and indeed was not part of the university. Management, stemming from the work of Frederick W. Taylor – an engineer – became part of the university even more recently[1]. Because they are idiosyncratic, universities differ from each other. For instance, some universities are prominent in veterinary medicine, others in dairy science, others yet in athletics. I do not believe Harvard would ever consider a department of dairy science. When I first came to this country, to the University of Minnesota, what perhaps impressed me the most was the existence of a program in mortuary science.

The university can do a number of things well. Obviously, it can teach. It gives value-added to students, even if, at times, it may reduce their creativity as they progress toward more advanced stages, from undergraduate to the Ph.D. The university also can do research and is today the major source, by far, of basic research in this country and many other countries.

At the same time, there are other things that the university does not know how to do and must learn to do. Unfortunately, the university does not know how to marshal its unique strengths to attack major social problems. For instance, in spite of being a great source of knowledge and education, the university has not been able to catalyze change in school education in this country to make it more effective. As another example, in spite of its great medical knowledge, the university has not been able to spearhead changes in health care. In this country we spend 12 percent of our gross national product for health care, yet our average life expectancy is not higher than that in countries that spend proportionately much less.

Neither has the university been able to do much about poverty. A recent *New York Times* article reported that those on welfare who are given a chance to graduate from college have a 99 percent chance of not falling back into welfare. Thus the university can do something about poverty – but certainly this is not enough. Much more is needed of the university in this as in all the other areas of major social problems.

To reiterate, the university in its collective body does have many of the skills and much of the expertise needed to attack these complex problems. But as an organization, it is simply not structured to address them. It is much better structured to address profound but simpler questions and opportunities. It can develop space telescopes; it can develop oceanographic ships; it can address new fundamental research themes. But it is not organized and prepared to address crucial social problems – education, health care, poverty, jobs, the environment – that are truly problems of the entire world.

Of course, nothing says the university should address these problems. Its wish to do so may be just one more example of its survival ability. Its lack of an adequate structure for an all-out commitment to address the problems may be just one more example of its supreme skill in avoiding being too precisely defined. That skill has enabled the university to survive as an institution for nearly 1,000 years. In a world

---

[1] Frederick W. Taylor (1856-1915) of the United States is considered the father of scientific management.

in which social priorities, sources of support and political systems change, an institution can better survive and thrive if it avoids pigeonholing and committing itself too strongly to a specific course of action.

## Response to Globalization

The university, being a cumbersome mechanism with arcane procedures, has long-term constants in its response to change, while many of today's problems require rapid change. The world is becoming rapidly globalized. Yet the university, in spite of the scholars and students it attracts from other parts of the world, remains in many respects a national institution. Industry has learned to operate globally, but the university has not yet learned to follow industry across national boundaries, even if it may have established a few centers or programs abroad. The university stays in place and hopes that it will attract students, industry and support, somewhat like a sea anemone waving its tentacles and hoping that nourishment will come to it, rather than like a deep sea swimmer roaming all over the ocean. Today there is not a single major university that is broadly supported by the global community. Telecommunications technology may change this.

As a political animal quite successful at survival during the millennium of its existence, the university has been basically unchallenged. Even the closing of Chinese universities during the Cultural Revolution was but a brief and ultimately insignificant episode. If anything, the university has been perhaps too successful in selling itself, to the point that it is tempting to say that the university itself has become an ideology. There is an almost universal belief in the importance of the university and a desire for involvement of the university in every significant issue that confronts our society. As an ideology, the university is a beneficial one. Among all the ideologies, it is the one that speaks to the rational component of humankind. It also speaks to universality and to the betterment of humankind through knowledge. Thus it is logical that the university would take relatively early to science. Only much later did it also take to technology, often through the creation of technological universities.

## The Technological University

The technological university is more purposeful than the general university, being, of course, more focused on science and technology.

In the United States, there are some 15 private – or as they are more properly called, independent – technological universities and a number of public ones. Some, such as California Institute of Technology, are more focused on science; some, such as Polytechnic, on engineering; and some, such as Massachusetts Institute of Technology, feel that they have graduated from being technological universities to becoming general universities. Finally, there are, of course, engineering schools within general universities, both public and private, which in the United States train the majority of engineering students. In Japan, the first department established at Tokyo University was engineering, thus challenging stereotyped models of the origins and evolution of the university.

The technical universities in this country could be viewed as harking back to the French Convention. In 1794 the Convention, inspired by the French Revolution and the "Siècle de Lumière," started the École Polytecnique in Paris. It took until 1834, when Rensselaer Polytechnic Institute (RPI) was established in Troy, New York, for the idea of an elite institution of science and technology to develop on this side of the Atlantic. The second institution of that kind after Rensselaer thirty years later in 1854 was Polytechnic. MIT came a few years later yet.

One important characteristic of most of these technological universities is that they became research oriented. It would be simplistic, however, to believe that the idea of an institution to train people in science and technology and to do research was unknown before Napoleon's time. Suffice it to think of Prince Henry the Navigator and his school of navigation in Sagres, Portugal, that was so important to global geographic discoveries – or, earlier yet, at the turn of the first millennium, of the Salerno medical school in Italy.

As institutions, the technological universities are only about 200 years old, but in the United States they have become significant beyond their numbers and size as strongholds and catalysts of technological education and research. Consider all that MIT has done – and yet it is not a large university, with perhaps no more than 10,000 students. Consider Cal Tech, one of the smallest, with fewer than 2,000 students. Together with the University of California at Los Angeles, it made the development of southern California possible through its research on high-voltage power transmission that enabled that region to draw power from the Sierras. Cal Tech also brought the aircraft industry to Southern California. Consider Carnegie Mellon: it contributed to changing the economy of Pittsburgh.

## Polytechnic University

When I came from Illinois in 1973, Polytechnic was in serious financial trouble and in turmoil as the product of a forced merger of two competitive entities – the Polytechnic Institute of Brooklyn and the New York University School of Engineering and Science. We had to decide what to do to survive, and we made the strategic decision that the future of Polytechnic lay in a close interaction with industry. But the problem was that, progressively, industry had left New York City, which had been the home of Polytechnic for more than 100 years. Industry had gone to Long Island; it had gone to Westchester, north of the city; it had gone to New Jersey, and it had gone to other parts of the country. So, first of all, we decided to follow industry within the limits allowed by our charter, that is, within New York State. On Long Island, to where some industry had moved (such as Sperry, which was started in Brooklyn), we revitalized a research and graduate center, transforming it into a full-fledged undergraduate and graduate campus for engineering. Next we established a graduate center in a different part of the metropolitan region – in Westchester – to be close to IBM, NYNEX and a number of other important companies in that area, southern Connecticut and northern New Jersey.

But still the major problem with which we were grappling was New York City itself. That was our historical base. What were we to do? What kind of industry was left in the city? What was the rationale for a technological university in the city? I remember visiting with a distinguished businessman who was one of the members of the Board of Regents, the board that supervises the state's education activities. He asked me, "What in the world is an engineering school doing in New York?" I said, "Please look around, everything you see is human-made. That's what an engineering school does in New York." But New York had lost its traditional industry. So we had to face the question of what the role of Polytechnic would be in the city. In the process of analyzing the situation, we discovered that while New York had lost much of the traditional industry, there was another kind of industry in New York – the financial industry – that was becoming a very heavy user of technology, but with which engineering schools had absolutely no familiarity. In engineering schools we always looked at the manufacturers of technology and never paid much attention, in terms of teaching or research, to the needs of the users of technology. I remember the sense of discovery when we learned of Securities Industry Automation Corporation (SIAC). SIAC is the technological nerve center of the financial industry in this country. It is not a developer of technology in any major sense; it is a sophisticated user of technology. It operates the telecommunication networks and the information systems for the New York and the American stock exchanges. Without SIAC, it would be impossible to have 300 million shares traded a day.

So we saw that there was in New York an industry that could hold in its technology the future of Polytechnic. We decided we would attempt to build a 16-acre university-industry park focused on this industry, called Metrotech. Metrotech will employ 15,000 people as well as operating a NYNEX-New York Telephone center that will train about 35,000 employees a year.

We first suggested the idea of a university-industry park in Brooklyn to the Board of Polytechnic in 1975; we are breaking ground only now. Thus, to build Metrotech has been a lengthy and difficult enterprise, particularly because we did not own the land and did not have the resources to acquire and develop it. We had to build coalitions with the city, the state, and industry. The city has committed itself to an investment of some $325 million, between infrastructure improvements, tax rebates and other incentives. Industry is investing about $1 billion in its own facilities. The state is financing a building for our Center for Advanced Technology in Telecommunication, which is the academic pivot of Metrotech. Alumni and friends invested in a new advanced library for Polytechnic – the Dibner Library – designed to closely interact with the telecommunications center. A comparison of Metrotech with Sophia Antipolis, the French research park on the Riviera, is instructive in showing the concentration or, if you like, the power of an urban research and industry park. Sophia Antipolis has 5,000 acres. Within its 16 acres, Metrotech has today a larger amount of commercial space already committed.

In developing Metrotech, we became ever more aware that the financial industry is a key industry for this country. I believe that the National Science Foundation should invest not only in manufacturing centers, but also in centers that are involved with the users of technology, such as the financial industry. Were the United States to lose its financial operations to Tokyo or London, it would lose one of the few industries that is a major source of its exports. The financial industry in Japan pays entry-level engineers higher salaries than the manufacturing industry – a further indication that the industry has become strongly technological and is beginning to attract some of the best brains among engineering students. In financial institutions one now encounters technologists at the vice president and senior vice president level, not to mention the chairman of Citicorp, John Reed.

To reinforce our interactions with industry in general, at Polytechnic we also joined forces with some distinguished alumni who are successful venture capitalists, to establish a venture capital fund, called Poly Ventures. New York is probably the venture capital center of the country, but that capital tends to be invested away from New York, especially on the West Coast and in New England. We said, "Why not in New York, to encourage the technological development in the area but not to the exclusion of promising opportunities elsewhere?" This was our goal, together with the desire to involve the expertise of as large a segment as possible of our faculty and, last but not least, to make money for the university. The funds for Poly Ventures – $30 million – came from the United States (particularly pension funds), and countries such as Taiwan, Singapore, Japan, and Switzerland. Unlike other technology venture capital funds involving universities, in ours the university as a whole is an active partner, rather than an individual faculty member or a group of faculty members.

The concept of an urban research park is now transcending Metrotech. An intriguing example is the Technopark being planned in Moscow, not too far from the Kremlin, as a partnership of an Italian company and a Soviet group to showcase Soviet research and encourage its commercialization in the West. The Technopark, somewhat analogously to Metrotech, will have a hotel, a conference center, and an office building, as well as research centers – one in biotechnology, one in materials, one in microelectronics and one in telecommunications.

Metrotech is also beginning to be an international educational magnet – a role that, I believe, holds great promise for the university. Polytechnic recently signed an agreement with an Italian telecommunications conglomerate to train a pool of Italian engineers who will come to Polytechnic for a master's degree in electrical engineering with a focus on telecommunications. They will spend one academic semester in Italy, taught in part by our faculty, and one at Polytechnic. The goal is to broaden the horizons of this group of engineers in a new environment – one made more unusual by Metrotech.

## Beyond Pragmatism: the Key Challenge

I have discussed the pragmatics of Polytechnic as a case history of a technological university and of university-industry interaction. But the technological university – any university – truly cannot be a university if it does not go beyond the pragmatics and address broader and deeper issues such as the cultural role of technology. As the university twists and turns to survive in an ever more demanding and rapidly changing society, it should not make survival an end to itself. Survival is the means to an end. Thus, at Polytechnic we have focused heavily also on the liberal arts, and we have created a center for philosophy and technology. At Harvard there is a great sense of the history and philosophy of science. The philosophy of technology, however, is very different. Whereas the philosophy of science ultimately revolves around the question of how we evaluate a scientific truth – and the implications of that truth – the philosophy of technology revolves around the question of the nature of an artifact. This is a different and most fundamental problem for a society like ours, which is constantly modifying nature and creating artifacts. Almost everything around us is an artifact – that is, human-made. The environment itself is becoming an artifact, and so are humans, thanks to implants, drugs and genetic manipulation. The fundamental question that the university should not be allowed to eschew is: "What is the nature of what we do, and why we do it?"

## Conclusion

Today the university is troubled in its purpose and needs to rethink itself. It needs to respond to new demands and, in doing so, to continue to assure its own survival. But it should not spend too much time in soul searching, as excessive introspection will cause us to miss the greatest possible opportunity – that of addressing immense and urgent social problems. It has been said that the best way to define oneself is through action. I believe that the universities at this moment must have the courage to do so.

# THE SWING OF THE PENDULUM: FINANCING OF BRITISH UNIVERSITIES FROM THE 1960s THROUGH THE 1980s

SHIRLEY WILLIAMS
*Public Service Professor of Electoral Politics*
*John F. Kennedy School of Government*
*Harvard University*
*Cambridge, MA 02138*

## Abstract

In the early 1960s, British universities were considered the guardians of liberal education. They were funded by the government through block grants; money for science was allocated through the Research Councils, which considered above all the reputation of the scientists. During the late 1960s through the 1980s, however, attitudes about universities and science changed, as did sources of funding, leading to more emphasis on applied research, as well as less freedom to pursue interests within the universities.

The pendulum always swings too far. As our perceptions change of what universities should be, of the objectives they should pursue, we tend to neglect their past accomplishments and even reject them. Yet universities exist in a long institutional timespan. Their sense of themselves develops organically. They can be persuaded, encouraged, coaxed into changing direction only gradually, like a supertanker or a tree. If they are forced suddenly into a new mold, the trunk snaps, and the institution begins to wither and die.

My ministerial life in the United Kingdom began in January 1966 as a junior minister in the Department of Labour. A year later I became a midlevel Minister of State in the Department of Education and Science in charge (under the Secretary of State) of higher education and the Research Councils, which allocated money for basic scientific research.

The British universities in the 1960s, disturbed a little by the first gusts of the student revolution, remained the guardians of a liberal education. Their educational objective was to instruct an elite based on intellectual merit rather than on position in society in subjects drawn from the classic curriculum. These subjects afforded an insight into human history, human knowledge, and human behavior: philosophy, literature, history, mathematics, and what was significantly still described as "pure science" – natural science rather than technology. Their research objective was the pursuit of excellence, financing the person rather than the project, because the person had demonstrated in his or her work qualities necessary to push outward the horizons of the subjects under study. The research undertaken was mostly cheap, interest-led research, because it was basic science, rather



*D. S. Zinberg (ed.), The Changing University*, 39–43.

than the development or the production of prototypes; interest-led because the scientists themselves determined the priorities through selection based on peer review.

What is sometimes called big science is not cheap even at the level of fundamental research. Subjects such as astronomy and molecular physics now use equipment so sophisticated that even substantial nations can no longer sensibly sustain their own research programs: witness the development of CERN as an international research center or the clubbing together of European Community countries to develop new optical telescopes or molecular biology laboratories. Nevertheless, in the 1960s, when I became responsible for the science budget, most Research Council money was distributed among scientists working in the life sciences, physics, or chemistry on relatively inexpensive projects, in laboratories established and maintained by the universities. This was the so-called "dual system"; the universities provided the capital, plant and equipment, and the Research Councils provided salaries and running expenses. The Research Councils avoided the expense of maintaining research institutes (there were and are only a few such institutes in the U.K.); the universities were able to add to the education of their predoctoral graduates and even some bright undergraduates by involving them in laboratories where outstanding scientists were exploring the frontiers of their subjects.

The research was primarily interest led; scientists themselves determined which application from their fellow scientists would be financed. Almost every proposed piece of research that was rated alpha in quality – alpha because of its presentation and often because of the reputations of those senior scientists associated with it – would be financed. The interests of industry and commerce went virtually unrepresented, apart from a few individual businessmen who served on the research councils and were not involved in selecting individual research projects. Indeed, few scientists in the U.K. at that time thought of the potential commercial value of their discoveries; information about some of the early research findings on monoclonal antibodies, for example, was innocently published by Medical Research Council scientists who wanted humanity to benefit from what they had found out.

The universities I knew in the 1960s in Britain were not atypical for Europe. In the older European universities, a liberal education combined with the pursuit of excellence in research constituted the model to which they aspired to conform. Indeed, technical universities and research institutes were founded precisely to meet the needs of industry and commerce that this hallowed model neither sought to satisfy nor succeeded in satisfying.

The complement to this tradition of detached, even lordly scholarship was a reputation for independence. Universities guarded the public interest and the values of free societies: open debate, dissent, a plurality of opinion. It would be wrong to underestimate their contribution in this respect since World War I, whether as opponents of the Nazis, of the satellite Communist regimes of Eastern Europe, or of apartheid in South Africa. The voices of academics were frequent in the chorus of opposition to the political pressures of intolerance and the political claims to monolithic truth.

It has been a significant role, still needed as democratic institutions emerge in Eastern Europe and South Africa and materialism pervades the West. We would be foolish to weaken the autonomy of the universities so that they can no longer fill this role. Amidst demands for better-targeted research with evident applications and for teaching to meet labor-market needs, voices calling for the right of free inquiry can be drowned out. Yet it is more essential for our free societies than any technological breakthrough.

Both in the United States and in the U.K., the system of block grants to universities between the wars and for a short time thereafter helped sustain interest-led research. In the U.S., foundations were willing to make substantial sums available for general research, in the hope of long-term, unidentified outcomes. In Britain, the money was public money, budgeted for higher education and science by the government, but paid out through the University Grants Committee (U.G.C.) or the Research Councils. The majority of the U.G.C.'s members were nominated by the universities, though formally appointed by the minister. The U.G.C. allocated the money among universities and occasionally indicated the desirability of expanding a particular subject. The universities' funds were then supplemented by grants from foundations or private sources, though those rarely amounted to more than 5 or 6 percent of their overall budgets.

The peer review system, by which the Research Council grants were determined, was little questioned. It was sufficient that the group thought well of its peer and his/her students. It was not easy for an unknown scientist, however gifted, to break into this caste system, its leaders ennobled by fellowship in the Royal Society. It was a system that inspired and encouraged apprentice scientists to gather in a team around the Great Man (or Woman). It was a system that, in its clear delineation between peer-group science (pure science, Royal Society science) and applied science (engineering and technology) relegated the latter to a lesser rank.

By the late 1960s, the classic university model was fading away. The universities had won crucial support from the egalitarian politicians who dominated the government then. The universities were willing to expand two or three times over to accommodate the bright children of the middle and working classes. They defended meritocracy over aristocracy and thereby won a new bipartisan consensus.

But even as the universities grew, so did their problems. Demonstrations and the occupation of university buildings by students soured public and political attitudes toward them. Taxpayers, who saw students as rebels and layabouts, resented the financial burden of student grants. In the rapidly expanding universities, often obliged to recruit faculty of only medium ability to meet the demand, the system of tenure sometimes protected poor teachers and indifferent scholars. The costs of research multiplied; the sophistication of scientific equipment drove its cost well ahead of inflation. In an effort to keep the good times going, scientists argued that investment in research and development was closely correlated with economic growth and competitiveness in testimony before parliamentary committees, by scientific columnists, and by politicians. It became received wisdom. But the argument was not borne out by the evidence. The United States spent proportionately more than any other country on basic research in the 1950s and the 1960s, with the U.K. next and France not far behind. But the countries that forged ahead economically were the Federal Republic of Germany, Italy, and Japan, at that time relatively low spenders and low achievers in basic scientific research.

To the increasing cost of research and the failure to demonstrate a close link between research and economic growth, I would add a third factor: the change in public opinion toward science. When scientific research had been relatively inexpensive and its findings little known or understood, public comment concentrated on its beneficial and "miraculous" achievements. The atomic bomb ushered in a new era, science as Cain as well as Abel, a force for evil and destruction as well as for good. But while the press argued about the consequences of the scientific revolution, they were still enraptured by its potential. Science seemed capable of anything. The Soviet launch of Sputnik in 1957 captured the public imagination. Putting a man on the moon became, for the United States, not so much a scientific imperative as a political one. President John F. Kennedy, according to one story, was approached by his science advisor, Jerome Wiesner. "Look, Mr. President," Wiesner said, "if we spend the same money on education in science and engineering as we plan to spend putting a man on the moon, we could transform our economic prospects." "Jerry," replied Kennedy, "you don't understand. I can get the money to put a man on the moon. I can't get it for education."

President Richard M. Nixon was another politician who understood that public opinion could shape priorities for research. He announced that his administration would wage war on cancer by financing every promising line of research. Lack of money would be no barrier. Nixon's war on cancer had this in common with Vietnam: it was flawed by belief in technological overkill, that given enough resources, science and technology could solve all problems and defeat all challenges.

The apparent failures of science and technology to win these and other wars or to be the engine of competitiveness and growth have fostered a climate of skepticism. The public is now ambivalent about such technological achievements as nuclear energy and nuclear fusion, fearful of the effects of pesticides and herbicides on the environment, unenthusiastic about medical advances that delay death only to prolong dying. Such skepticism has been nourished by arguments among scientists, once a united and silent priesthood, about the effects of radioactivity or the likelihood of nuclear fusion replacing nuclear fission as a source of energy. No one knows whom to believe anymore.

Basic science would henceforth have to prove its usefulness by pointing to relatively short-term material results. The most effective way to concentrate on specific outcomes was to alter the sources of funding. Industry became more significant as a source, the Research Councils less so. In the United States, foundations became unwilling to make block grants and instead tied their grants to specific outcomes. The accent in research moved from basic to applied, in response to the feeling that excellence in basic research had not yet yielded market leadership, because of a failure to develop and apply its findings.

The Research Councils themselves became more commercially aware; no longer did they idealistically inform the world of each new research breakthrough. Scientists were enjoined to patent their discoveries and to seek industrial partners to exploit them. The emphasis in the Research Councils' budgets moved toward engineering and pharmaceuticals, more likely to succeed commercially than particle physics or tropical diseases. Scientific research became the handmaiden of industry and commerce.

The traditions of liberal education and student choice became controversial, too. Industry was short of qualified scientists and engineers as its products became more sophisticated. In the late 1960s and 1970s, the British government advocated reserving two-thirds of the new places in higher education for students of science and engineering. In labor planning terms, the policy made sense. The shortage was self-evident; the jobs for graduates were available. But the school leavers rejected careers in applied science and engineering, choosing the social sciences instead. The 1960s generation had become disillusioned with the effects of science and technology, and the promise of jobs was not enough to persuade them to take up careers as scientists and engineers.

In the 1980s these early winds of change in higher education intensified. The Margaret Thatcher administrations favored polytechnics, with their concentration on teaching applied science and technology, over the universities. Polytechnic student numbers expanded faster than numbers in the universities. The University Grants Committee, dominated by representatives of the universities, was replaced by the University Finance Committee (UFC), which was directly under the control of the Secretary of State and had a preponderance of businesspeople as members. New powers enabled the UFC to remove funding from specific departments of universities, and consequently a number were closed down, among them departments of philosophy, classics, and the like. Tenure was abolished for any new appointment, though it was retained by those who were neither promoted nor changed jobs. The universities faced a difficult choice: to become much more dependent on private sources of money, including fees from overseas students (and probably, soon, fees from home students also), or to accept the intervention of a government bent on making them respond to industrial and commercial needs in exchange for public funds.

This revolutionary change in the funding and administration of higher education in Britain carries its own costs, and they are incalculable because it is hard to measure what has been lost and what has been gained. Gained are a greater sensitivity to the market, a keener awareness of what can be commercially exploited, and a more entrepreneurial attitude among academics. Lost is the climate of free inquiry and co-operative endeavor in which the scientific imagination flourishes. Unquestionably the quality of basic science has suffered seriously. Lost, too, is the role of universities as guardians of debate, deliberation, and dissent, for such activities engender controversy.

The great Russian historian Solovyev once said that liberty flourished in the interstices of the state, like grass in the cracks of pavements. The same might be said of creativity. Those who want to harness the universities to commercial objectives may destroy the very qualities they admire in them – intellectual excellence, free inquiry, scientific imagination. The pendulum has swung too far.

# CHANGING PATTERNS OF FINANCE FOR HIGHER EDUCATION: IMPLICATIONS FOR THE EDUCATION OF SCIENTISTS AND ENGINEERS

MAUREEN WOODHALL
*Centre for Higher Education Studies*
*Institute of Education*
*University of London*
*58-59 Gordon Square*
*London WC1 England*

## Abstract

Throughout the industrialized world universities are adapting to changes in the level and sources of funding; universities in the U.K are examined in detail. Competitive forces will increasingly determine the income of universities. Adaptations include a greater emphasis on income generation and cost accounting, fewer tenured staff, and a shift from basic to applied research.

## Introduction

Throughout the developed world there have been significant changes in the past decade in the level and sources of funding for higher education, and in the financial mechanisms used to distribute funds to universities and other higher education institutions. A recent study for the Organisation for Economic Cooperation and Development (OECD), *Changing Patterns of Finance in Higher Education*, examined trends and changing patterns of finance in eleven OECD countries and concluded:

> It is only recently that policy makers have begun to understand the critical links between expenditure patterns, costs and efficiency in higher education and the mechanisms by which institutions receive funds....Many governments now see financial incentives as a more effective way of influencing the pattern of activities in higher education institutions than administrative intervention."[1]

---

[1] Organisation for Economic Co-operation and Development (OECD), *Financing Higher Education: Current Patterns* (Paris: OECD, 1990). For a summary of the study see G. Williams, "Changing Patterns of Finance" in *The OECD Observer*, January 1990.



*D. S. Zinberg (ed.), The Changing University*, 45–53.

This paper draws on the OECD study and the eleven country reports on which it is based, and a study of changing sources and mechanisms of funding higher education in the United Kingdom based on a series of 24 case studies of British universities, polytechnics and colleges in progress at the Centre for Higher Education Studies of the Institute of Education in the University of London. It summarizes some of their main findings, and considers the implications of these changes for the education of scientists and engineers.

## The OECD Study

A series of country case studies was written in 1988, covering eleven OECD countries: Denmark, Finland, France, Germany, Greece, Japan, the Netherlands, Norway, Spain, the United Kingdom and the United States. The ways in which universities are funded in these countries differ considerably. For example, private universities, which charge tuition fees, are common in Japan and the U.S. but not in most European countries. Systems of student support also vary significantly, although increasingly student aid is made up of a mixture of grants and loans. Several countries have recently introduced loans, or increased the proportion of student aid provided in the form of a loan.

The study found three basic ways of funding higher education institutions: unconditional grants and subsidies from public funds; conditional grants, which restrict the way funds may be used; and market mechanisms, which provide income for universities from the sale of academic services. The study concluded that "The political climate of the 1980s in several OECD member countries is leading to a shift away from unconditional public funding towards conditional grants and the market....Public funding agencies are becoming increasingly selective; in several countries they are beginning to see their role as 'buying' academic services on behalf of the community rather than managing or regulating institutions."

The report notes various changes under way or under discussion in member countries including: more sophisticated formulas to determine allocation of funds to institutions; greater financial autonomy for the institutions once they have received funding; increased proportion of income from student fees; a sharper distinction between funding of research and teaching; a trend toward "competitive bidding" by institutions for public funds; and an increased proportion of university income derived from contracts with industry and commercial organizations.

Increased reliance on formula funding, which directly links a university's grant with the number and level of its students, the type of courses offered, and other factors, have developed in Denmark, the Netherlands and the U.K. Several countries are trying to reduce public funding for higher education and increase the proportion of funds from industry and commerce and from students and their families. Private funding is particularly important in Japan and the U.S. but many other countries are currently trying to increase private contributions. For example, Australia has introduced a "higher education contribution," collected by means of a graduate tax, and the British government has introduced full-cost fees for overseas students.

Higher education institutions and policy analysts have expressed a number of concerns regarding these changes in funding patterns, including the danger that applied research and peripheral activities such as short courses will prosper at the expense of basic research and core teaching; growing disparities among institutions and between subject areas, with prestige institutions benefiting at the expense of others, and with science and engineering benefiting at the expense of humanities, languages and arts; secrecy and problems of intellectual property rights arising from commercial sponsorship of research, which may restrict academic freedom and may slow down the progress of scientific discovery by impeding the full flow of information; and too much academic time devoted to income generation and bidding for funds at the expense of time available for teaching and research.

Advocates of market mechanisms on the other hand believe that increased competition for funds and more selective allocation will improve efficiency and make institutions more responsive to changing labor market needs and new scientific and technological developments.

Many of these changes in higher education funding have implications for science and technology. The trend toward selectivity may benefit science and engineering, as governments seek to target funds on priority areas, but at the same time the rising costs of scientific research may mean that competitive bidding will threaten high-quality research programs. More reliance on the market may lead to a shift of resources toward science and technology, in response to economic pressures, or it may threaten high-cost subject areas and lead to the neglect of basic research, for which no commercial sponsors can be found.

Some of these issues have been explored in more detail in research on monitoring and evaluating new funding mechanisms in higher education that is currently being conducted by the Centre for Higher Education Studies at London University.

## Changing Patterns of Finance in British Universities

Since 1980 there has been a decline in the proportion of university income derived from government grants, and an increase in the proportion from non-government sources. At the same time funding has become more selective, with a shift toward specific research grants and contracts, rather than unconditional grants.

Figures 1 and 2 show trends in university income in Great Britain from government and non-government sources between 1981 and 1987. The proportion coming from general, unconditional exchequer grants has fallen from 64 percent of all university income in 1981-82 to 56 percent in 1987-88, while the proportion derived from overseas student fees, research contracts, and consultancy with U.K. industry, business and commerce rose from 5 percent to 15 percent.

Several universities now derive less than half their total income from general exchequer grants and more than half from fees, research grants and contracts and payment for services, including consultancy. Many of the universities that have been most successful in increasing their share of private funding are those with a strong scientific and technological reputation, with close links to employers and to industry and commerce.

A survey of business funding of higher education found a wide variety of financial contributions, including financial sponsorship of students and academic staff; payment for short courses for employees; research grants and contracts; consultancy; endowments and gifts (including equipment); and joint ventures for the development of commercially viable discoveries.

The value of industrial grants and contracts for university research more than doubled between 1982 and 1988, and total income from industry and commerce represented about 7 percent of all university income in 1987-88, although in individual universities the proportion ranged from 0.2 to 10.5 percent. Subject areas differ substantially in their capacity to generate income. Medicine and engineering, followed by physical science, attract the highest proportion of income from industry, and humanities and law the lowest.

These changes in sources of finance have had a significant effect on university management and organization, time distribution of academic and administrative staff, and the development of different academic areas. The main organizational changes reported include the creation or strengthening of senior management posts concerned directly or indirectly with income generation, including the promotion of industry links, overseas student recruitment, fund raising, public relations and institutional research management; an increase in the size and influence of finance offices, and a shift of emphasis toward greater financial control; growing concern with determining the costs of all institutional activities; the appointment of senior staff, including not only vice chancellors and directors, but also heads of departments, schools and faculties, who have senior management experience outside higher education; strengthening lay participation in the government of the institution, particularly on planning committees; the establishment of income-sharing arrangements at the departmental or faculty level to encourage cost saving and income generation; increasing use of performance indicators to compare institutions and individual cost-centers within institutions; and an increase in the proportion of staff on short-term, rather than permanent contracts, and a reduction in the number of tenured staff.

The effects of these changes on the academic work of universities is difficult to assess, but many of the institutions we visited reported such changes as the development of modular courses, special courses for particular groups of students such as overseas students, more flexible arrangements for credit transfer, increased emphasis on attracting research contracts and consultancies, and a growing proportion of academic staff time devoted to income generation. Some also reported problems such as unbalanced age distribution of staff and limited opportunities for young people to progress.

Course modularization and credit transfer are usually justified on academic grounds but several institutions are promoting them as ways of increasing the size of teaching groups and the range of courses that can be offered. Special courses which charge full-cost fees have been developed for overseas students and employers who want specific courses for their employees.

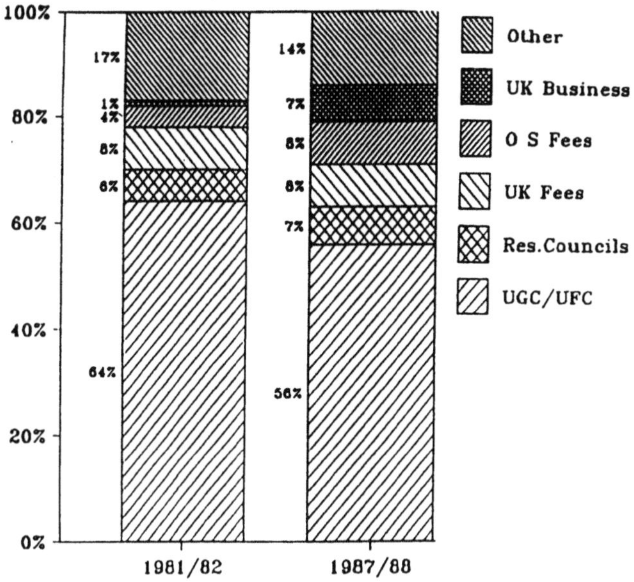


*Figure 1.* Percentage sources of university funds, 1981-82 to 1987-88 (United Kingdom).
*Source*: Williams et al. (forthcoming)

There has been a shift in the balance of research, away from pure research, funded through general research grants and concerned primarily with the advancement of knowledge, toward more applied research, funded through contracts and concerned with the use of existing knowledge to solve specific problems. There has been some concern that excessive growth of applied research may cause a decline in the rate at which new knowledge is created. There is no clear evidence that this point has been reached but there are widespread claims that the seed corn of human capital is being used up in problem solving rather than knowledge creation.

In virtually all the universities in our sample, staff reported that they now had to spend more time generating income through research contracts, consultancy and short courses. The decision to give greater priority to income generation has led many institutions to appoint new staff with marketing skills and to develop institutional strategies designed for local, national or even international markets.

Particular attention has been paid to the overseas student market since the introduction of full-cost fees for overseas students in 1980. The immediate effect was a sharp fall in the number of overseas students coming to Britain. The total number of overseas students in universities, polytechnics and publicly funded colleges fell from 88,000 in 1979 to 55,400 in 1983, but since 1983 the number has been rising. By 1988 there were 72,000 overseas students in higher education in Britain, of which 49,000 were in universities, nearly 25 percent more than before full-cost fees were introduced. This recovery reflects

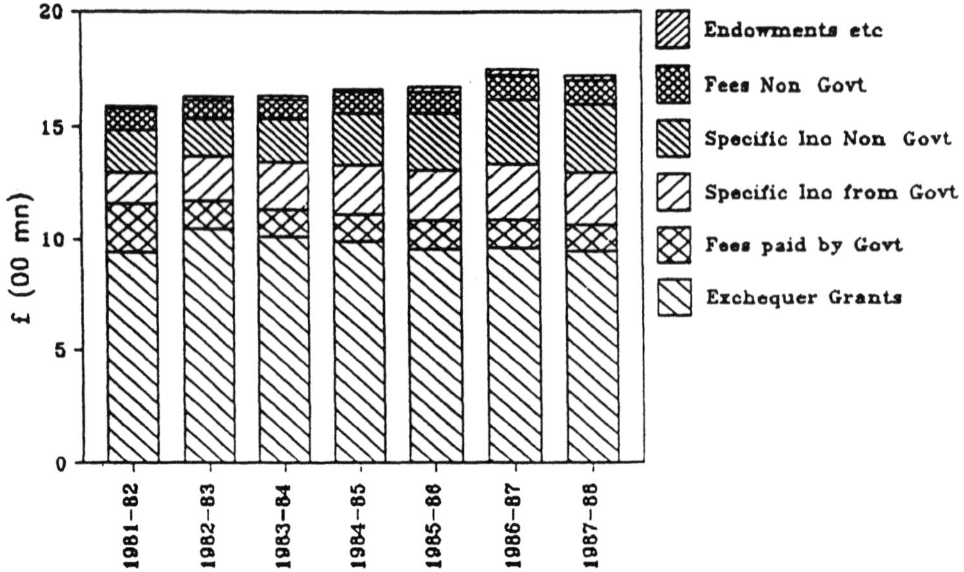


*Figure 2.* University funding from government and non-government sources, 1981-82 to 1987-88 (United Kingdom).
*Source*: Williams et al. (forthcoming)

both an increase in government support for overseas students through targeted scholarships (22,000 were assisted in 1988-89) and an increase in recruitment activities by universities, including the introduction of new courses, improvements in academic and welfare provision, as well as more active marketing of higher education overseas.

## Implications for the Education of Scientists and Engineers

One of the aims of government policy has been to increase the supply of scientists and engineers through more selective funding, by means of initiatives such as the Engineering Technology Programme (ETP) designed to shift funds and resources to engineering departments and thus increase the supply of graduate engineers. Other initiatives included the Alvey Programme, designed to promote collaborative research in information technology, through joint university-industry projects, and a number of other programs designed to encourage closer links between universities and industry.

Several of these programs have had a positive impact on departments of science and engineering but there have been criticisms of the short-lived nature of several of the initiatives and the high opportunity costs of "competitive bidding" for funds, which requires considerable academic staff time, often at the expense of teaching or research.

Not all the new funding mechanisms have favored science and engineering. The introduction of full-cost fees for overseas students differentiated between arts, science and medicine, so that students of science and engineering have to pay fees about one-third higher than those studying arts, languages, social studies or business. Since full-cost fees were first charged in 1980, the proportion of overseas university students studying science and engineering has fallen from 51 percent in 1978-79 to 41 percent in 1987-88. The most popular subject area is now social, administrative and business studies, which accounted for nearly 30 percent of all overseas students in 1987-88, compared with 20 percent in 1978-79.

Government-funded scholarships target certain countries or categories of students but there is no specific program for scientists and engineers. A recent review of government policy on financial support for overseas students listed the main objectives of scholarship programs: to win influential friends overseas by enabling future leaders, decisionmakers and opinion formers from all walks of life to study in the U.K.; to help the development of skills and human resources in developing countries; to promote the security and prosperity of the U.K. by cultivating good political and commercial relations with other countries; and to ensure a continued supply of world-class research students in U.K. universities.

## Future Funding Patterns

Future patterns of funding for universities will be determined by two government policies that are already evident and that the government intends to continue to implement. There will be a shift to fees from exchequer grants as sources of income. The government has announced that from 1990 home student fees will be increased from about £600 a year to £1600, and the funds allocated by the Universities Funding Council (UFC) will be correspondingly reduced, even though for the present all home students will continue to have their fees paid in full from public funds. In the future, fees are likely to constitute an increasing proportion of university income, and differential fees for different subjects will be introduced for home students as well as overseas students. This will mean that a university's income will be even more dependent than at present on the number of students it can recruit, particularly the number studying science and engineering, who will bring higher fee income than arts and humanities students.

The second trend is toward increased selectivity in research funding, which was introduced by the Universities Funding Council in 1989. In the future a university's research income will increasingly be determined by its previous success in generating research income from the Research Councils, business and other sources of research grants and contracts.

Together, these two trends mean that universities will have to devote more time to attracting students, both home and overseas, and to generating research income than in the past, when university income was dependent on formula funding. The government believes that this increased reliance on market mechanisms to determine higher education funding will increase the efficiency and accountability of institutions.

**Conclusions**

The OECD study concludes:

> A funding mechanism is not merely a device for allocating resources from providers to users. It is a system of control and an important two-way channel of communication between the providers and users of finance. The terms on which funds are offered show the priorities of those who supply them; the ways in which they are used reveal the preferences of those who receive them.[2]

There are instances of tension or even conflict between the priorities of providers and users of funds. The priorities of government funding do not always match the priorities of employers. For example, government policy makers may emphasize the need for more scientists and engineers at a time when accountancy and business studies appear to offer a higher private rate of return. In the same way, the priorities of commercial sponsors of research may not be consistent with the priorities of university researchers. For example, industry values secrecy, while university researchers favor the free exchange of information.

Nevertheless, there are signs that changing patterns of finance for higher education are achieving some policy goals, for example an increase in income derived from industry and commerce, and the development of closer links between universities and industry.

Our research on the effects of changing patterns of finance is still continuing and more detailed findings will be published in 1991.[3] Already the system of finance is undergoing further changes. From 1990-91 the Universities Funding Council will allocate a larger proportion of research funds on the basis of judgments about the quality of research, and tuition fees will rise. Both the Universities Funding Council and the Polytechnic and Colleges Funding Council (PCFC) now require institutions to bid for funds on a competitive basis.

These changes raise a number of important questions. The shift toward market mechanisms and more reliance on client-led research and courses raises the question of who is the client? Is it the student, or industry, commerce and other employers, or is it some notion of the national interest, however defined? Another important issue is the ownership of knowledge, in conditions in which an increasing proportion of research is funded by industry, commerce and private agencies, which may favor secrecy at the expense of open dissemination of knowledge among scholars.

The fact that market and competitive forces will increasingly determine the income of universities also raises the question of whether competition is an appropriate model for funding higher education, and what should be the basis for competition. The introduction

---

[2] Williams, op. cit.

[3] G. Williams, et al., *New Funding Mechanisms in Higher Education* (Forthcoming, 1991) and G. Williams and C.P.J. Loder, *Business Funding of Higher Education* (Forthcoming, 1991).

of competitive bidding for students, as well as research contracts, is intended to reduce the costs of higher education and to make institutions more responsive to the needs of the labor market and the economy, but critics fear that it will reduce quality. Questions remain about whether institutions will choose to compete in terms of price or quality, and how quality can be measured.[4]

The changes in sources and mechanisms of funding that have occurred or have been planned mean that there are likely to be major changes in higher education institutions both in the U.K. and in other countries. The search for funding mechanisms that encourage efficiency, accountability and responsiveness to the changing needs of society while safeguarding academic quality and autonomy is likely to continue well into the next century.

---

[4] See C.P.J. Loder, *Quality Assurance and Accountability in Higher Education* (London: Kogan Page, 1990).

# CONTRADICTIONS AND COMPLEXITY: INTERNATIONAL COMPARISONS IN THE TRAINING OF FOREIGN SCIENTISTS AND ENGINEERS[1]

DOROTHY S. ZINBERG
*Program for Science, Technology and Public Policy*
*John F. Kennedy School of Government*
*Harvard University*
*Cambridge, MA 02138 USA*

## Abstract

The education of foreign scientists and engineers (S&Es) has emerged as a significant policy issue as host countries debate how many should be educated and at what cost, and sender countries such as China struggle with problems of brain drain and political insubordination. In general, the mobility of foreign S&Es has worked well, reinforcing the international openness of science, creating an expert work force, contributing to diversity and knowledge, among other benefits. However, information has come to be equated with property and S&Es themselves as a form of intellectual capital. Research universities now more tightly coupled to industry or in some instances, the military, are being asked to justify their policies concerning the education of foreign S&Es. This paper examines the practices and policies of five industrialized countries in an increasingly complex policy area.

## Introduction

For years the education of foreign science and engineering students (S&Es) was a subject that attracted only modest national or international attention, usually under the rubric of foreign students in general. But recently this topic has come to dominate discussions as diverse as the economics of education, the ethics of brain drain, the responsibilities of industrialized to developing countries, human rights, economic competitiveness, national security, foreign policy and international relations – all in addition to universities' traditional concerns about admissions policies, appropriate curricula, and funding.

The complexity has evolved from the growing interdependence of economics, education, technology and national security – the relative weights of which are as yet unclear. For example, should foreign policy take precedence over university autonomy in determining which and how many foreign students are accepted to a university?

---

[1] Study supported by the National Science Foundation, No. SRS-8609985, "Training Foreign Scientists and Engineers: Goals, Dilemmas and Achievements."



*D. S. Zinberg (ed.), The Changing University*, 55–87.

Should foreign students be allowed to immigrate to the host country in defiance of their home governments? Should they be expected to pay full fees, or should their educations be viewed as aid to developing countries? Should Eastern European students studying in NATO countries, for example, be allowed access to technology that is banned for export to their home countries?

During the past five years nations have become more aware that their long-term economic and military security is based less on natural resources than on human resources – particularly scientists and engineers. Consequently, they and the knowledge they produce have become valuable commodities, sought after in the marketplace among universities, industries and governments. Scientists and engineers themselves have come to embody the technological revolution. They move readily across international boundaries, catalyzing the transfer of information while propelling the inevitable globalization of science and eventually that of the work force.

Though deemed vital for the well-being of scientific inquiry, the international mobility of S&Es often generates an uncomfortable contradiction between maintaining a national identity and reaping the rewards of globalization.

Since World War II universities that seek foreign students and faculty members have increased the momentum for nations to become global players. Entering into institutional arrangements that abet internationalism, many universities have established branches abroad, such as Stanford's Kyoto campus. Others have set up cooperative arrangements with foreign universities; the University of Salford in the United Kingdom, for example, eager to break down "the walls between island Britain and the rest of the world," sends chemistry students for one year to the Grande Ecole in Lyons, France (after intensive training in French), or to the University of Toledo in the United States. At the (West) Berlin Technical University, the science and engineering departments offer "sandwich degrees," a nation-wide program for scientists and engineers initiated by the Deutscher Akademischer Austauschdienst (DAAD), a program that develops and coordinates a wide variety of scholarship and exchange programs. Students from third-world countries begin their Ph.D. studies at home, come to Berlin for research experience and then return home to complete the degree.[2] Or some universities have set up inter-university relationships such as that maintained by the University of Marburg in Germany and Moscow State University in the Soviet Union, which have exchanged students throughout the Cold War.

University research laboratories have established links via computers, dedicated telephones and even yearly migrations of staff and graduate students from one country to another, particularly among industrialized countries. Some of this mobility involves multinational corporations. For example, the chairman of Cambridge University's Department of Computer Sciences in Britain spends three months each year at Digital Equipment Corporation (DEC) in Palo Alto, California. Twelve of his former Ph.D. students live within five miles and continue to work with him there. DEC also contributes

[2] Interview with Professors Kunkel, Gumlich, Naunin and Janositz at the Berlin Technical University, FRG, January 23, 1987.

to the support of his laboratory in Britain and funds three U.S. graduate students to work with him at Cambridge.[3]

The majority of these arrangements work well, but as the line between basic and applied research has become indistinct, questions have arisen. Many leading molecular-biology centers in Europe and the United States, subscribing to the ideals of open communication in science and the belief that the products of science should benefit all societies eagerly accept and train scientists from the People's Republic of China (PRC). Of course, these centers recognize that they gain considerably from the contributions of talented foreign students. However, if these Western-trained Chinese scientists return to the PRC and launch biotechnology companies, unconstrained by the patent and trade laws adhered to in the West, and then compete with or even preempt a Western product, the molecular biology centers find themselves torn between their ideals and their national interests.[4] Countries with leading scientific enterprises face the dilemma of wanting to retain their leadership and to strengthen their own economies, while also wanting to contribute to international science, the education of foreign scientists and the industrialization of a developing country.

These contradictions have generated tensions, largely unanticipated, which have begun to cast a shadow over an important worldwide activity, the education of foreign S&Es. As the associate provost and vice-president for research at a major U.S. research university said:

> Foreign S&Es will become an issue for every research university. We're caught in a conflict between the university's former role as an ambassador and its current role of generating strength for the economy. We have to balance pluralistic interest groups.
>
> The policy of former years was pro-international.... let's Americanize the world. But it's no longer appropriate. There's lots of pressure building up in Washington. Should we really be providing education on this scale for foreign S&Es?[5]

This perspective, which is disputed emotionally by many colleagues in the same university, is one indicator of the intensity of the debate. The issues are intricately bound up with the very principles of free exchange that underlie the governance of universities as well as the democratic societies in which they exist. However, these principles are coming into conflict with local, national and international priorities.

The study reported here grew out of the recognition that the education of foreign S&Es will increasingly be dominated by national economic and foreign-policy considerations in industrialized countries. This study of foreign S&Es (1986-89) examined the policies of

---

[3] Interview with Professor R.M. Needham, chairman, Department of Computer Sciences, Cambridge University, U.K., June 6, 1987.

[4] Interviews, Europe, 1986.

[5] Confidential interview at a U.S. research university, October 1986.

five countries that are committed to strengthening their existing programs and also to grappling with concern about economic and military security: the Federal Republic of Germany (FRG), France, Japan, the United Kingdom (U.K.) and the United States (U.S.).

## Chinese S&Es in the U.S.: Tension Made Visible

Chinese students abroad have provided the most dramatic and current example of the social and political consequences of the education of foreign students, particularly S&Es. Since the brutal massacre of University of Beijing students in Tiananmen Square shocked the world in the spring of 1989, the fate of Chinese students abroad has elicited public and political attention.

During the days before and after the events at Tiananmen Square, PRC students in the U.S. and Europe could be seen nightly on TV telefaxing material, orchestrating large banks of computers and broadcasting via satellites to their peers in Beijing and elsewhere. Relaying information from city to city in the PRC or even within Beijing, where students were without access to each other, they passed along Western TV journalists' predictions of how the PRC government would behave and kept up telephone communication with the families and friends of the Beijing students. The technologies that many of these students were using in their laboratory work in the States showed people around the world how the information-technology revolution had made it possible for foreign nationals to communicate instantly across the globe despite the wishes of their government. The vivid pictures demonstrated that the PRC government was helpless to control the spread of essential information.

Once the government moved in to crush the students in Beijing, the fate of those overseas became problematic. Technology also worked against them: Chinese officials monitored U.S. and European TV programs to learn which PRC students abroad were sympathetic to the democracy movement. Concerned about reprisals, these students sought to postpone their return to the PRC.[6]

Even before the birth of the democracy movement, the Chinese government in 1987 had begun to take drastic steps to bring home students, especially those S&Es in the U.S. they believed vital to the country's development. Parents in China received letters insisting they tell their children to return; work units threatened to fire parents of students who did not. In several U.S. cities PRC consulate officials approached university administrators, asking them to force students to return to China, and made it difficult or impossible for students to obtain the Chinese documents required to extend their visas.[7]

---

[6] A report issued by the Committee on Scholarly Communication with the People's Republic of China (National Academy of Sciences), as quoted by Betty Vetter, "Problems for Chinese Students in the U.S.," *The American Association for the Advancement of Science Observer* 3 (March 6, 1989): 10.

[7] Dorothy S. Zinberg, "Perspective: PRC Science Students and Scholars Abroad," *Science* 239 (March 25, 1988): 1,475-6.

At that time the U.S. government assisted the Chinese government by refusing visa extensions for some 5,000 Chinese who had earned degrees in the U.S.[8] A State Department official claimed that the Chinese students had become "an irritant to the bilateral relationship and a problem for U.S. universities because they linger in the financial system and take up a lot of staff time to deal with their limbo status." He added, "Underfunding [of the students on Chinese government stipends] creates frustration for the students and reduces the effectiveness of the educational experience."[9]

Distrustful of the country's intellectuals, the Chinese government has been torn between its wish to modernize, which would require many well-trained scientists and engineers, and its awareness that some would not return and those who did might be difficult to control. The government had not anticipated that the majority of S&E students in the U.S. at that time did not plan to return.

The first students sent abroad in the late 1970s were older and thus tied to their families and workplaces at home. Almost all returned to China. The students in the next wave, beginning in the mid-1980s, were younger and more independent. When large numbers of them chose to remain abroad, the government was embarrassed politically. It responded to the Tiananmen Square uprising by further intensifying its efforts to bring students home and tightening the requirements for going abroad. The U.S. has been targeted in part because of the large numbers of Chinese students in the country – more than 75 percent of all overseas PRC students and more than four times as many as in the next largest host country, the U.K., which had 5,600 in 1988 (see Appendix A). In addition, before the uprising, other countries had made permanent residence difficult to obtain, so that a number of students had come to the U.S. from host countries such as the U.K. and West Germany where they had been students. (After the uprising the host countries extended the students' visas and provided stipends when they were no longer forthcoming from the PRC.)

Of the approximately 40,000 Chinese students in the U.S. (no one knows precisely how many there are), 29,000 arrived in 1988.[10] Beginning about seven or eight years ago most major research universities and institutes in the U.S. have come to rely heavily on Chinese research and teaching assistants in science and engineering and do not want to lose them or discourage others who might apply.[11]

The movement to protect these students against deportation has become a complex

---

[8] Vetter, op. cit., p. 10.

[9] Department of State official at a meeting of the China Committee of the National Academy of Sciences, January 16, 1988.

[10] Vetter, op. cit., p. 10. The State Department has issued about 56,000 visas to Chinese students and scholars from 1979 through 1987.

[11] "IIE Publishes Findings of U.S. University Survey," *China Exchange News* 17 (December 1989): 18. See also "Chinese Students Abroad: A Growing Interest," *China Exchange News* 16 (3): 2-7; and Linda Moussouris, "A Study of Harvard University International Students in the Sciences and in Engineering," August 1989, NSF No. SRS86-0985.

political contest in the U.S. – easily the most visible foreign-student incident in this country since the Iranian hostage crisis a decade ago (1979-81), when the U.S. Immigration and Naturalization Service admitted with embarrassment to Congress that it did not know the whereabouts of most of the Iranian students in the United States. In early 1990 President Bush and the Congress battled over legislation to provide special immigration status to PRC students. The tensions stemmed from the desire not to affront the Chinese government versus the desire to satisfy constituents' outcry over human rights.

The PRC students' plight commanded international attention. University administrators and faculty, human-rights and public-interest groups, politicians, China scholars and the Chinese students themselves testified in front of Congress, lobbied the president, organized public forums on campuses across the country and appeared on TV to argue for the proposed legislation.

In 1989 Congress voted almost unanimously for the Chinese Immigration Act, which would have given PRC students the right to remain in the country for two more years. Bush, however, vetoed the legislation, arguing that an executive order extending student visas for the same period – in effect, the president's promise – would suffice to protect the students until they were guaranteed safety at home. Legislation would provoke the Chinese government to further restrict student exchanges, he argued, and would threaten the U.S. military alliance with China and the more than 5,000 science and technology agreements in effect between the PRC and U.S. businesses; his "promise" would not.

The president prevailed, but the Chinese government imposed restrictions anyway. The new rules that took effect February 10, 1990, make it virtually impossible for university graduates to study abroad unless they promise to work for five years after graduation in the PRC. However, by June 1990 there were indications that new graduate students in the natural sciences were able to avoid this requirement.[12]

Alarmed by the support PRC students abroad gave to the democracy movement, a Chinese teacher stated, "The government does not want to help educate the gravediggers of socialism."[13] Yet the government knows that were it to cut off foreign S&E education completely or discourage foreign students and scholars from visiting China, the resulting intellectual isolation would doom its development. So it continues to allow some international mobility, even if at a sharply reduced rate.

It is too early to tell whether the new restrictions will be carried out. PRC ukases and actions are often not related, because large loopholes are built into edicts. Government-sponsored students (approximately 17 percent) would not be affected by the new rules, and government officials' children and individuals with political influence or money from relatives abroad have been able to ignore such pronouncements in the past.

---

[12] Glenn Shive, "Update: The Five Year Rule and its Implications," *China Exchange News* 18 (June 1990): 17.

[13] Sheryl WuDunn, "China Acts to Restrict Study Abroad," the *New York Times*, February 25, 1990, p. 3.

After the last round of restrictions on study abroad (1988-89), the numbers of PRC students in the U.S. actually increased 15.4 percent, from 25,170 in 1988 to 29,040 in 1989.[14] (PRC students have continued to apply to U.S. universities; those who had procured passports prior to February 1990 were able to travel abroad. The fate of more recent applicants has not been documented at the present time.)

The Chinese students' predicament poignantly underlines the urgency of examining national policies affecting foreign S&Es. Approximately 80 percent of PRC students – as compared with about 50 percent of students from all foreign countries – are concentrated in the physical sciences (particularly physics), mathematics, computer sciences, life/health sciences, and engineering: the fields most relevant to economic and military security and, consequently, those receiving most attention from university, industry and government worldwide.[15]

## Background: The Role of Demography

To comprehend the scale of the growing worldwide flux of foreign S&Es, it is helpful to examine the demographic trends of the international foreign student population. Since the end of World War II industrialized countries have welcomed foreign students. Japan has only recently begun to recruit foreign students, but the other nations in this project have been receiving and recruiting them vigorously for the past four decades. The overall numbers, which have grown 10 times since 1950, grew most rapidly between 1970 and 1979 (see Appendix B).

The growth of foreign students in the 1970s was in large part due to students going abroad from OPEC countries. On this basis it was predicted in 1980 that more than one million foreign students would be enrolled in the U.S. by 1990.[16] Even though the growth has slowed – some 360,000 foreign students were in the U.S. and more than one million worldwide in 1989 – the numbers have become significant politically, economically, socially and educationally.

In the U.S. six countries – China, Taiwan, Japan, India, South Korea and Malaysia – now provide more than 40 percent of all foreign students. Forty-five percent are in graduate schools, 36 percent in bachelor degree programs, and the rest are divided among associate-degree, language-training and non-degree programs (see Appendices C and D).

The impact of this infusion can be readily seen in the large contributions they are

---

[14] Marianthi Zikopoulos, ed., *Open Doors 1988-89: Report on International Educational Exchange* (New York: Institute of International Education, 1989), 23.

[15] Leo Orleans, *Chinese Students in America: Policies, Issues and Numbers* (Washington, D.C.: National Academy Press, 1988); quoted in Vetter, op. cit., p. 10.

[16] *Foreign Students and Institutional Policy: Toward an Agenda for Action* (Washington, D.C.: American Council on Education, 1982), 2.

making. In 1988 foreign S&Es accounted for more than a quarter of all Ph.D. recipients.[17] Foreign S&E students now hold 66 percent of the postdoctoral positions in engineering, 55 percent in the physical sciences and 42 percent in the life sciences.[18] In 1988-89, 50 percent of all graduate engineering students and 40 percent in math and computer science were nonnationals. Since 1986 the entire increase in graduate enrollments in science and engineering – about 9,000 students – has come from non-U.S. citizens.[19]

In Europe and Japan the distribution is more difficult to calculate because the degree structure varies from country to country. However, in France, the U.K., and Japan approximately one-third of the foreign students are in graduate programs and, with the exception of France, are concentrated largely in engineering and science. For example, in the U.K. 52.4 percent of all foreign graduate enrollments are in science and engineering (see Appendix E). The FRG makes no distinction between undergraduate and graduate degree programs so that it is difficult to draw comparisons.

In 1988 the National Research Council, an agency of the National Academy of Sciences, National Academy of Engineering, and the Institute of Medicine, questioned whether it was wise to rely on the immigration of doctoral scientists (almost 60 percent of assistant professors in U.S. schools of engineering who are under the age of 35 are nonnationals), what their impact might be on higher education, and how they might affect the future supply of U.S. citizens in science and engineering.

Their report raised the question of whether the presence of so many foreign-born engineering professors and teaching assistants might discourage U.S. students from entering the field because "teaching roles place stress on communications skills and cultural acclimation -- the areas where foreigners are likely to be weakest."[20] In addition, they noted that the working relationship between universities with large numbers of foreign S&Es and laboratories that engage in national defense or industrial competitive research could be impaired.[21] While strongly affirming the U.S. commitment to educating nonnationals, the council pointed out that these issues had the potential to contradict the country's commitment to free scientific exchange and increase the tensions between different national interests, as in the case of Chinese S&Es in the U.S.. It stressed the opportunities the continued inflow of foreign S&Es provided, but also the potential problems that could occur, were the numbers of Americans studying engineering to continue to decline.[22] The subject has continued to gain national and international

---

[17] *Science and Engineering Indicators*, National Science Board (Washington, D.C.: U.S. Government Printing Office, 1989), 55.

[18] Ibid., p. 54.

[19] Ibid., p. 225.

[20] *Foreign and Foreign-Born Engineers in the United States: Infusing Talent, Raising Issues* (Washington, D.C.: National Academy Press, 1988).

[21] Ibid., p. 162.

[22] Ibid., p. 162.

attention.[23]

Because France and the FRG allow almost no changes from student status to permanent residency, citizenship or professorial appointment for S&Es from outside the European Economic Community, these concerns are absent. In addition, because both France and the FRG ascribe high status and salary to engineers, the most underpopulated academic field among the sciences in the U.S., they do not suffer from shortages. The U.K. allows some leakage in understaffed fields, primarily engineering, while Japan allows virtually no shift from student status.

Thus it appears that the U.S., particularly with the pending immigration reforms that would enable more foreign S&Es to obtain permanent visas, is moving in a different direction from the other countries in the study. All of them want to attract talented foreign S&E students; only the U.S. wants them to remain -- a position that might encounter resistance from the country's S&Es if defense funding continues to drop and large-scale layoffs ensue.

## The Economic and Political Context

Part of the increased attention paid to foreign S&Es in recent years stems from the rapid changes in international economic and political conditions. For one, oil-producing countries lost considerable influence and spending power as oil prices plummeted. Between 1970 and 1975 the number of Saudi Arabian students in the U.S. increased almost fourfold – from 1,404 to 4,026; during the next decade the number quadrupled once more to 17,283, only to decline with the loss of oil revenues.[24] Suffering from political repression, war and a ravaged economy, Iran went from being the leading source of foreign students in the U.S. in 1979-80 – 17.9 percent of the total, with 51,310 students – to 2.4 percent of the total with 8,950 students in 1988-89. Like most sender countries, Saudi Arabia and Iran's students are heavily concentrated in engineering and technical fields.[25] For the eighth consecutive year, Middle Eastern enrollments shrank substantially in 1989 – by 3,500, the largest numerical decline in one year for any world region.[26]

Eurosclerosis -- the fear that Europe had dried up culturally, economically and politically in the early 1980s -- has given way to Europhoria. The Japanese and West German economies have burgeoned, and not surprisingly, so have the numbers of Japanese and Germans abroad. Although the U.S. has become the world's largest debtor, its role as the world's leading educator has not been adversely affected. However, the

---

[23] For a discussion of U.S. demography, see Dorothy S. Zinberg, "The Next Generation of Engineers: A New Breed?" *Aspen Institute Quarterly* 2 (Spring 1990): 50-57.

[24] Zikopoulos, op. cit., p. 8; UNESCO figures.

[25] Ibid., p. 21.

[26] Ibid., p. 19.

"new" Europe aims to compete directly in education within the next decade, and Japan aspires to take on the mantle for the Pacific Rim countries. The speed of political and social change is nowhere more evident than in the aftermath of the crumbling of the Berlin Wall in November 1989. Within days hundreds of East German S&Es were awarded grants to West German universities, Westerners were welcome in Eastern European universities, and the international mobility of Soviet S&Es surged. Interestingly, few predicted the onset of these radical shifts. The speed with which Europe has been achieving economic integration, particularly in cooperative science and technology projects, was slow to be recognized. Hardly had the recognition occurred, when the rapid downfall of communism in Eastern Europe altered the balance of power. Now, once more, it is difficult to predict what will develop next, as the economic impact of the unification of Germany – and its technical work force – is uncertain in the near term although likely to be a positive force for the economy in the long term.

Many individuals interviewed for this study strongly stressed the need to rethink their respective countries' policies – or, in the case of the U.S., lack of policies – regarding foreign S&Es. Some in every country stated that this should be done in the context of stepping up the recruitment of foreign S&Es. In universities dependent on revenues from foreign students, especially those with undersubscribed engineering departments that otherwise might be closed by budget-strapped administrators, faculty energetically subscribe to pro-recruitment policies. Even in the FRG and France, where students do not pay tuition, the overall emphasis is on recruiting students, although both countries have recently tightened entrance requirements. In the FRG the numbers of foreign students have risen sharply in the last two years from 75,000 to 91,000. Many universities are overcrowded and do not recruit; others with underpopulated engineering departments try to attract qualified students. And the DAAD has negotiated special arrangements with the governments of Indonesia, Tunisia, and Iraq, which send large numbers of their students to study engineering primarily as well as other subjects at their governments' expense. The FRG enrolled more than 20,000 students from East Germany in the fall of 1990, although they believe that the movement of students from west to east will not occur for a number of years even after unification because of the poor conditions in East German universities and cities.

Those individuals in favor of increasing enrollments support the traditional values underlying the openness of the university and the mobility of S&Es; they also believe the very presence of foreign S&Es will help propel the host university and country into the global scientific enterprise, which they consider inevitable. A less-than-active exchange of students, they argue, would deprive the countries of the knowledge needed to compete globally.

At the same time, in most industrial countries the universities have strong ties to industry and government research. The U.S. is the only country in the study that has important, readily discernable links to the military. Although the U.S. Department of Defense provides only eight percent of research and development funds for universities, 32 percent of the funds go to engineering departments (50 percent of electrical engineering and 42 percent of aeronautical/aerospace R&D) and are concentrated in approximately ten universities.[27]

University research often leads to commercial applications and militarily-useful technology. The British prime minister's science adviser, John Fairclough, in an interview reported in a British newspaper, stated that universities could no longer determine their own research agendas; rather, he argued, industry should set them because it better understands the nation's needs.[28]

Industry, especially high-technology companies, feels threatened by many of the countries from which the foreign S&Es originate. As a result, ambivalence and outright xenophobia have surfaced in some sectors of academia, government, and industry. A French official commented that he wanted to increase intra-European and U.S. exchanges of S&Es but cut back drastically on the numbers of third-world students. "France," he said, "cannot afford to give away its national patrimony."[29] Such intense assessments surrounding foreign S&Es are apparent in the titles of recent reports, such as: *Fondness and Frustration*, *Boon or Bane* and *Obligation or Opportunity*.[30]

The enthusiasm for recruitment – as heralded in the title, *Open Doors*, a publication of the Institute of International Education – contrasts sharply with the ambivalence and sometimes strident xenophobia of some academics and policy makers. These contradictions should be explored internationally so that the traditional benefits of international mobility are not eroded. Unfortunately, these contradictions are not easily quantifiable. Yet they must be probed in order to assess the real and imagined threats to the vitality of the scientific enterprise.

For these and other reasons, it is necessary to understand what social, educational, political, economic and security issues are involved, as universities and governments begin to examine their recruitment policies for foreign S&Es.

## Success: What Works Well

Overall the international mobility of S&Es is a success story. As the sustained growth in the numbers of S&Es studying abroad attests, students, universities, host and sender countries alike are benefiting. Training foreign S&Es serves many diverse purposes.

---

[27] Figures provided by Dr. F. Karl Willenbrock, assistant director for Scientific, Technological, and International Affairs, National Science Foundation, December 1989.

[28] Quote from the *Independent*, discussed at a meeting at the British Council in Paris, January 20, 1987, by Professor William Mitchell, director of U.K. Science and Engineering Research Council.

[29] Interview with official at the National Center for Scientific Research (CNRS), Paris, January 1987.

[30] Crauford D. Goodwin and Michael Nacht, *Fondness and Frustration: The Impact of American Higher Education on Foreign Students with Special Reference to the Case of Brazil*, Institute for International Education Report No. 5 (1982); Elinor Barber and Robert P. Morgan, *Boon or Bane: Foreign Graduate Students in U.S. Engineering Programs*, IIE Research Report No. 15 (1988); Alice Chandler, *Obligation or Opportunity: Foreign Student Policy in Six Major Receiving Countries*, IIE Research Report No. 18 (1989).

While providing students with an education that is often unobtainable in their home countries, the process also strengthens science in the host country and creates an expert work force for the sender countries, the host countries or the international work force. Furthermore, foreign S&Es contribute to diversity in universities where students have little experience with foreign languages and cultures.

Talented students also bring scholarships and other awards to departments. An electrical engineering professor at Cambridge University (U.K.) volunteered, "It would be a pretty parlous state without overseas students. Our research budget would drop by 30 percent."[31]

The intellectual contributions of foreign S&Es, particularly at the graduate and postgraduate level, are reflected in the numbers of coauthored papers they write and their candidacy for faculty positions, particularly in the U.S. Internationally coauthored scientific papers also have risen dramatically. In Canada, France, the FRG and the U.K., one-third to one-half of coauthored publications are written with colleagues from other countries. In engineering and physics combined with technological subjects worldwide, the leap in international collaboration has been astounding – up 58 and 48 percent respectively between 1973 and 1984.[32] In the bioengineering department at the University of Texas, the top three candidates for the two appointments to assistant professorships in 1989 were Asians with foreign citizenship. The university has recognized the importance of foreign students and young faculty by initiating "a very good language and culture course to help them integrate into the life of the university," according to the department chairman, an enthusiastic supporter of foreign students at the university.[33]

In the U.S. and the U.K. foreign S&Es bring in tuition as undergraduates and support departments – such as engineering, mathematics, chemistry, agriculture and public health – that have been shrinking as a result of changes in demography and academic popularity. Without them, many departments would be seriously understaffed, lacking teaching and research assistants. In London a tropical-medicine department has only foreign students, and at the Technical University in (West) Berlin 60 percent of the students in the metallurgy department are foreign. Although the official FRG national policy limits foreign students to eight percent in the entering class of all departments, the administration is able to accept more when FRG nationals do not apply in adequate numbers.

Most international cost/benefit studies have argued that the overall benefits of educating foreign S&Es outweigh any monetary costs. In one way or another, the students pay for their own educations. When foreign engineering students, for example, have received their undergraduate educations at the expense of their own governments and families, then

---

[31] Interview with Professor Alex Broers, Cambridge University, June 9, 1987. A substantial portion of the funds for foreign S&E graduate education are provided by British Trust Funds and scholarships.

[32] *Science and Engineering Indicators*, National Science Board (Washington, D.C.: U.S. Government Printing Office, 1987), p. 98.

[33] Interview with Professor Henry Bose, Department of Bioengineering, University of Texas-Austin, March 23, 1989.

remain in the host country after obtaining a master's or doctorate degree, they contribute to the science and engineering work force as teachers, researchers, industrial scientists or as professionals in related occupations. If they return home or migrate to a third country, they can then become goodwill ambassadors for their former hosts, continuing to speak the host country's language and thereby enhancing its international status. The FRG, sensitive to the decline in the use of the German language internationally (a situation that appears to be reversing itself with German unification and the strength of the FRG economy) has stated explicitly its goal to increase the use of the language by educating foreign S&Es.[34]

Foreign S&Es who have returned home or migrated to a third country often collaborate with former faculty and colleagues, thereby increasing the intellectual productivity of both countries. Some economists argue that they also contribute to the host country's economy by buying its export products. A British economist believes that overseas students should be publicly financed, because hosting them is a form of "export promotion that will stimulate economic growth and employment."[35] A prime example of this is Jordan, where the Minister of Energy purchased the solar hot-water heaters now widely used throughout the country from a U.S. manufacturer, a former classmate at Texas Agriculture and Mining University (A&M).[36]

For multinational corporations, foreign S&Es are well-suited for employment in their subsidiary corporations around the world. Siemens, IBM, British Petroleum and Schlumberger, for example, all hire foreign graduates of their respective countries' universities to staff their operations overseas.

In addition, educating foreign S&Es is considered charitable, the initial post-World War II impetus for accepting students from war-ravaged countries and more recently from newly industrializing and third-world countries.

Most important, the education of foreign S&Es reinforces the international openness of science, a precondition for its vitality.

## Contradictions: What is Not Working Well

Many of the assumptions about the benefits of educating foreign S&Es are being challenged when different institutions within a host country set mutually exclusive goals. One of the most striking contradictions arises when a country is striving to be a major force in international science and engineering, yet at the same time wants to maintain

---

[34] Interview with Dr. Karl Roellofs, general secretary of the Deutscher Akademischer Austauschdienst,(DAAD) West Berlin, January 25, 1987.

[35] Robin Marris, "Assessing the Commercial Element in the Overseas Student Policy," *Readings in Overseas Student Policy*, ed. Gareth Williams, Martin Kenyon and Lynn Williams (London: Overseas Student Trust, 1987), 7-86.

[36] Personal communication, Amman, Jordan, March 20, 1979.

national identity and autonomy.

The majority of U.S. scientific leaders and government officials are convinced of the need to be part of the growing internationalization of science. Differences of opinion exist only concerning the means to achieve this goal. One indicator of the support for internationalization is evident in the increased numbers of academic exchange visas granted by the U.S. government, up 52 percent since 1978.[37]

Despite the growing internationalism or perhaps, in part, in response to it, many countries are experiencing rekindled nationalism in their ambition to be first in scientific R&D. The tension in industrialized countries between the concerted effort to be preeminent in "national" science and the inexorable movement toward global science is most evident in applied (or potentially applicable) fields such as molecular biology, where a revolution promises so much for agriculture and medicine, or in condensed-matter physics and electrical engineering, with their demonstrated achievements in supercomputers, semiconductors and superconductivity.

Although policy differences among countries are distinct, they are often less pronounced than differences among agencies and institutions within a given country. Federal immigration and tax authorities, state legislators and public university administrators, deans and department heads, as well as alumni and faculty all struggle to enact policies that are often mutually contradictory.

## Nationalism Versus Globalism

The tension between nationalism and internationalism as it relates to the education of foreign S&Es is clearly evident in two countries that come from different ends of the spectrum. Japan, until recently a closed society, is being pushed by successive prime ministers to internationalize the student body and by other countries to grant access to the laboratories and industries its students are given abroad. However, Japanese Department of Education officials and some faculty speaking off the record register serious misgivings about the new direction for which they believe the country and particularly the universities are unprepared.

In the U.S. the internal tension is in the other direction, with a significant number of officials in universities, government, and industry starting to question the wisdom of accepting large numbers of foreign S&Es. Some express concern that the presence of foreign S&Es could discourage industry from investing in university R&D because they fear trade secrets may be given away. Externally, both industrialized (the U.K. and to a lesser extent the FRG) and third-world countries (especially in Asia) have accused the U.S. of creating a brain drain of enormous proportions. Some faculty, even those who have "no concern that the students are taking away secrets," do worry that a predominantly Asian faculty could radically alter the university's culture in the future.

---

[37] *Science and Engineering Indicators*, op. cit., p. 98.

These concerns have led to questions why young Americans are not more interested in science and engineering and how to increase their recruitment.[38] In every country in the study, but particularly in the U.S., educators expressed concern about the declining interest of young nationals in science and engineering careers. Many in the U.S. agreed that the problem was not that there were too many foreign students, but that there were too few Americans.

*A. Japan*

Under the leadership of former Prime Minister Nakasone, the government in 1983 produced a White Paper that announced plans to increase annual foreign student enrollment from 10,000 to 100,000 by the turn of the century. This foreign-policy initiative was part of a larger effort to internationalize trade and business. *Kokusaika* (internationalization) quickly caught on.[39] However, Department of Education bureaucrats resented the fact that educational directives were coming from the prime minister's domain and dragged their heels.[40] A minority of outspoken Japanese academics were equally unenthusiastic.[41]

The obstacles were innumerable, the dissenters argued. Primarily an exporter of students – some 30,000 to the U.S. in 1989-90 – Japan was ill prepared to host students from abroad, they said. Certainly the language and culture presented significant barriers for outsiders.

Because of its large deficits, Japan has placed ceilings on all government expenditures except for foreign students, the military and international technical assistance. The annual increase over budget has continued at 15 percent, enough to maintain the recruitment program; however, the Japanese government pays only 10 percent of foreign-student costs. The rest is expected to come from foreign governments, foundations and private funding.

In 1984 critics pronounced the project doomed. But by 1986 the numbers had begun to grow, and by 1988 more than 25,000 foreign students were in Japan. Among the graduate students 56.1 percent were enrolled in science, agriculture, medical sciences and engineering – 31.1 percent in engineering alone.[42] At the Tokyo Institute of Technology and Kyushu University, for example, 43 percent and 42 percent, respectively, of doctoral-level foreign graduate students are studying engineering.

---

[38] Interview with Professor William Mack Grady, Department of Computer Engineering, University of Texas-Austin, March 23, 1989.

[39] "Country Report: Japan," from the Organisation for Economic Co-operation and Development Seminar on Higher Education and the Flow of Foreign Students, Hiroshima, Japan, November 8-10, 1988, p.2.

[40] Interview with Dr. Peter DeAngelis, National Science Foundation representative in Tokyo, September 1986.

[41] Interview with Professor Kazayuki Kitamura, October 14, 1988, Cambridge, MA.

[42] "Country Report: Japan," op. cit., p. 16.

> This phenomenon [engineering enrollments] has won the attention of the media and has aroused great concern especially among a few elite and established universities. This is a trend which is being followed with great interest and concern by the government and the universities involved.[43]

The official policy still aims for 100,000 by the year 2000, although most prognosticators now believe 75,000 is more likely. At least one administrator still believes the goal of 100,000 foreign students is not unrealistic. "Rather," the director of the Institute for Higher Education at Hiroshima University says, "it is a necessary goal in the aim of democratizing and internationalizing the country."[44] The Japanese, he adds, have not yet dealt with the questions of the long-term employment and immigration opportunities for former S&Es. At present some graduates take part-time entry-level jobs to accumulate yen and then leave the country. It has been almost impossible for them to find good jobs in industry in Japan. However, Japanese companies are beginning to change. Whereas they have in the past hired only Japanese graduates of Japanese universities, they now need people who speak English well and are able to negotiate in the international world.[45]

In the meantime four national universities – Tokyo, Kyoto, Saitama and the Tokyo Institute of Technology – want to be prepared for future growth in the population of foreign S&Es, who are not expected to be fluent in Japanese. So they are experimenting with the use of English in technical courses. However, 80 percent of Japan's foreign students come from Asia – China, Korea and Taiwan – and the majority of those who decide to study in Japan speak Japanese but have no knowledge of English, which is needed to study mathematics, science and engineering.[46] Those Asians who speak English but no Japanese, such as many Chinese students, generally prefer to go to the U.S.

Feared and resented by other industrialized nations because of its closed-door business policies at home and scavenger policies abroad, Japan plans to become more accessible to foreigners through the exchange of students. It is the country's hope that the attitudes in more established host countries will soften. A recent study of more than 100 Japanese universities found that 62 percent would like to have more foreign students. However, some universities are reluctant to receive any more foreign S&Es, particularly those in large urban areas where foreign students are already concentrated.[47]

---

[43] Ibid., p. 16.

[44] Kitamura interview.

[45] Kitamura interview.

[46] Kitamura interview.

[47] "Country Report: Japan," op. cit., p. 16.

*B. The United States*

Unlike Japan in its effort to cope with ascendancy, the U.S. is struggling to deal with the fact that it is no longer first in all fields of science, engineering and high technology. For several decades following World War II the U.S., unravaged by war and the recipient of much of the scientific brainpower forced from Europe, was able to build a formidable empire. As long as government funding grew and the country's leadership in science and engineering was unassailed, the concerns about the numbers of foreign S&Es were small ones, and there was little urgency to consider a policy for foreign students. In 1970 a group of organizations that kept watch on the growing numbers of foreign students recommended:

> Each university should develop an explicit rationale for the admission of foreign students and prepare itself for closer scrutiny by boards of trustees or regents, as well as by state and other funding agencies, as to why these students are being admitted and supported.... There is a need for a long-range national policy on international exchange of graduate students to which individual institutions and graduate schools can relate their own policies.[48]

Little attention was paid to this recommendation primarily because in the U.S. departments select their own graduate students, and few professors in science or engineering were likely to have read a foreign-student policy publication. Colleges and universities continued to attract and recruit foreign S&Es enthusiastically. Visas were readily available, and departments – most often individual professors – continued to (and still do) decide which foreign graduate and postdoctoral students to admit without consultation with a central administration. The issue that did attract attention had less to do with foreign S&Es, but rather with "bounty hunters" who recruited foreign students for colleges with declining enrollments and funds or with specific problems encountered by foreign S&Es because of inadequate funding or social isolation.

But in the early 1980s new questions were being asked about foreign S&Es, and attitudes toward them began to shift – not among their teachers but among a small number of industrialists and government officials. The U.S began losing market shares in major technology industries, and the federal government concluded that military security was being threatened by the presence of foreign S&Es, particularly in high-technology fields. Concerned that the U.S. was "hemorrhaging technology" to the Soviet Union, the Department of Defense during the Reagan administration requested some university

---

[48] College Entrance Examination Board (1971), as quoted by Y.G.-M. Lulat, Philip G. Altbach, David H. Kelly, *Governmental and Institutional Policies on Foreign Students: Evaluation and Bibliography* (Buffalo, N.Y.: Comparative Education Center, State University, 1986), 64.

officials to conduct covert surveillance of foreign visitors and to limit their activities.[49] Even small leaks of technical information from U.S. universities to the Soviet Union were greeted with alarm in Washington. The contention by the National Academy of Sciences that the U.S. would lose more than it could gain by government agencies trying to oversee universities was greeted with skepticism.[50]

In 1989 John Deutch, provost of the Massachusetts Institute of Technology (MIT), succinctly outlined what he deems to be the dilemmas faced by his university and others with the high proportions of foreign S&Es.[51] MIT enrolls the highest percentage of foreign students in a U.S. university in relation to total enrollment (20.9 percent).[52] In 1988-89 there were 1,981 foreign students, up 5.3 percent from 1987-88.[53]

Deutch pointed out that the principle that had guided the admission of foreign students since World War II were based on assumptions that were "shattering in front of us, in fact, [are] shattered." He enumerated the previous assumptions:

> 1. We would educate Eastern Europeans. This has not occurred in any significant numbers.[54]
> 2. We'll educate third-world people – from Mexico, the Philippines, Venezuela – and they will go back and help build their countries. But they say we're stealing them because we need them – no more sending them back; they're a source of labor. Now there is no more helping the LDCs [less developed countries], and the immigration laws are very confused.
> 3. We must help the West Germans and the Japanese rebuild. Now they are [seen as] our mortal enemies. We would rather get DRAMs or tritium from the Russians than the Japanese. The whole notion of alliance, of commonality on education, where we think we have a comparative advantage of ideas, has completely gone by the boards. You cannot imagine some of the hatred I hear or the intense criticism I get personally because MIT permits Japanese companies to be part of our industrial-liaison program, because we permit Japanese students to get electrical-engineering degrees, or [when I]

---

[49] Personal communication; and Ross Gelbspan, "When Scientists Get Aid from the U.S.," the *Boston Globe*, January 23, 1984, p. 1.; and Richard Higgins, "Boston Colleges Seen as Targets of Espionage," the *Boston Globe*, October 17, 1988, p. 1.

[50] *Scientific Communication and National Security*, National Academy of Sciences (NAS), also referred to as The Corson Report, September 1982; see also John Shattuck, "National Security Controls in the United States: Implications for International Academic Science and Technology," in this volume.

[51] Address to Dual-Use Seminar of the Science, Technology and Public Policy Program of Harvard University, February 1989.

[52] Zikopoulos, op. cit., p. 62.

[53] Ibid, p. 103.

[54] No Eastern European countries are among the top 64 sender countries to the U.S. The figures were similar in France, the U.K. and the FRG, but in the year since his remarks were made, the situation in Eastern Europe has changed drastically. The FRG, the U.K. and the U.S. have been receiving a steady flow of students from Eastern Europe.

try to buy a piece of Japanese equipment.
4. We need the best minds, but they're not [foreign students]. There is an inverse correlation between the number of foreign graduate students and quality. [Many departments and schools seek graduate students to meet the needs of research grants and contracts, Deutch later added. This need sometimes dominates quality considerations and leads to accepting a higher proportion of foreign students.]
5. A relatively honest congressmen now says, 'I want this money spent on American citizens.' That kind of issue is going to become much more prevalent and is going to become very serious for MIT. And leaders from industry say [to MIT], 'You can't take money from Honda.'[55] Or you will be asked by an old friend at the University of Rochester, where a program is supported by Eastman Kodak, to enroll a student who works for Fuji in Japan, because Eastman Kodak does not want him in 'their' program. The University of Rochester retracted its position after many scores of scientists voiced loud disapproval, but the point about the growing encroachment of industry on university autonomy was made.

Deutch ended his remarks by saying that the issue of foreign S&Es is not broadly understood. His remarks were not his own views, but rather issues he had heard raised in the national debate and not adequately answered by the university community, he later said. "The basic principles were probably right," he added, "but now we need some real policies. We need a fundamental re-examination of the role of foreign S&Es in the U.S. today." Many academics present at the speech vehemently disagreed with the criticisms being raised about open U.S. universities, even though there was general agreement on the need for a re-examination of policies. Several seminar participants argued that universities must accept students according to their talents, regardless of their citizenship – a sentiment summed up by a professor who observed, "If you [U.S. universities] deviate from [this] principle, you'll have a second-rate university."[56]

Unpalatable as many of the issues raised by Deutch may be, they are opinions that have been put forth frequently in Europe as well as in the U.S. No country is exempt from these tensions. At a seminar in London a British academic dismissed as "rubbish" the notion of future foreign trade as motivation for educating foreign S&Es. Industrial espionage is well under way at universities, he argued, and the open-door policy needs rethinking. The director of foreign students in a West German university complained that for many of his colleagues, foreign students (who were mostly in engineering) were too much of a drain on their time, the university's facilities, and the country's finances. He hoped government policy would change to raise admissions criteria and further restrict admissions. And a French professor of physics resented the presence of so many Africans

---

[55] Subsequent congressional hearings chaired by Rep. Ted Weiss substantiated Deutch's statement that the government is opposed to Japanese and other foreign investments in U.S. universities.

[56] Joseph S. Nye, professor of government and associate dean of international affairs, Harvard University.

in his classes.

The major differences in Europe are due to the absence of military research in universities (especially abhorred by the Germans and the French), the free tuition provided by France and the FRG and the numbers of foreign S&Es. The U.S. educates more foreign S&Es than all the other countries in the industrialized world combined. Calculating the real costs in every country has proven difficult because of unquantifiable variables beyond economic ones. In the same way, it has been largely impossible to measure the social, scientific, educational and political gains or losses in educating foreign S&Es. Yet increasingly, legislators and administrators ask the question, How much does it cost to educate foreign students?

### Who Should Pay For the Education of Foreign S&Es?

The prevailing ethos in all four of the Western countries in this study is that all academically qualified S&Es should be eligible for admission. The practice, however, is different. One important indicator of how a country values foreign students is in how it handles tuition fees, because funding can effectively determine who can study. In the FRG funding of foreign students is viewed as a form of foreign aid for third-world countries. In addition, the FRG tries to encourage its own students to study in third-world countries. A recent experiment in the U.K., on the other hand, attempted to place more of the burden on students and their countries, and the U.S. is considering policies with this aim in mind as well.

#### *A. Federal Republic of Germany*

Of the five countries in this study, Germany has the most centralized and comprehensive policies for foreign students. At every level – federal, state (Lander), and university – the policies are coordinated through a number of quasi-governmental or private agencies that try to increase the numbers and quality of foreign students. To date, they have been effective. For example, the DAAD oversees numerous exchanges and fellowship programs, and the Alexander von Humboldt Foundation actively seeks accomplished scientists internationally to study and carry out research in the FRG.[57] There are no fees for foreign students, although the government has begun to insist that students prove they can pay for living as there are almost no job opportunities. (Determining the numbers of foreign students in the FRG for comparison with other countries is complicated, because government figures of foreign-student enrollments include guest workers' children, who

---

[57] See Charles V. Kidd, ed., *Proceedings from Seminar on Mobility of S&Es between the Federal Republic of Germany and the U.S.* (Washington, D.C.: American Association for the Advancement of Science, June 1987); Gert Fieguth, "Foreign S&Es: An Overview of Students in the FRG" (unpublished: 1989); and Chandler, op. cit.

were born and educated in the FRG and would be considered citizens in the U.S. Some 30 percent of the official foreign-student count – some thirty percent of whom are Turkish, Greek, and Iranian students – fall into this category.[58]) The cost for the government is approximately $25,000 per year per student, almost 10 times greater than that spent by the U.K. and twice that by France.[59] Although individual faculty members occasionally express resentment and foreign students have reported discriminatory remarks, the FRG is committed to the education of foreign students, and there has been no discussion of repealing the subsidy.[60]

*B. The United Kingdom*

Britain provides an instructive example of the unanticipated consequences of introducing full-cost fees. Starting in 1981, the government stopped subsidizing foreign students (except for European Community nationals, refugees and reciprocal-exchange students) and increased fees by 300 percent. The results were dramatic: the number of foreign S&Es, which had increased by 150 percent from 1970 to 1978, fell by 50 percent in science and 20 percent in engineering within a year. For the first time the overseas-student question was pushed to the forefront of political and educational debate.[61]

For many science and engineering departments, already reeling from budget cuts, the loss of revenue was severe. The Imperial College of Science and Technology, one of Britain's top universities, lost an additional 13 percent of its income. Students from the Commonwealth countries decreased by 40 percent between 1979-84.[62] Malaysia, which had been one of the major Commonwealth sender countries, retaliated with a "Buy British Last" campaign that cost British companies an estimated 500 million pounds in lost sales.

The British government tried to recover with a plan to provide scholarships to the best-qualified third-world students. The funding, known as the Pym Plan, came from the Foreign Office budget, an acknowledgement that education of overseas students is linked to Britain's economic and diplomatic well-being.

Universities also adapted by becoming more entrepreneurial. Foreign recruitment

---

[58] For a discussion of FRG foreign-student population, see Fieguth, op. cit.

[59] Chandler, op, cit.

[60] Since the interviews for this study were conducted (1986-89), rapid political changes in Eastern Europe and the imminent unification of the two Germanys have produced a totally new set of factors, which will undoubtedly affect the recruitment of overseas students. Part of the drive for foreign S&Es stemmed from the drop in the number of university-age FRG students. With the influx of German nationals from East Germany and other former German states already fluent in the language, it is likely that universities and technical schools will become crowded, and the active recruitment of students from third-world countries will decrease.

[61] Lynn Williams, "Overseas Students in the United Kingdom: Some Recent Developments," *Readings in Overseas Student Policy*, op cit., p. 13-24.

[62] Ibid., p. 13-24.

became a way to balance budgets, since universities were allowed to retain income from overseas students' fees. Imperial College rewarded any faculty member who recruited an overseas student with a round-the-world air ticket. At Salford University energetic recruitment combined with a close relationship with industry led to innovative new S&E programs, and enrollments zoomed. Salford's four-year courses in chemistry or engineering combined with European studies, including a year of carefully supervised industry placement in a European country, have proven highly successful.

A U.K. survey showed that after the institution of full fees, 70 percent of universities and polytechnics developed entrepreneurial activities: advertising, promotional visits abroad by academics, and links with foreign universities.[63] Science faculties at Oxbridge colleges have resisted the trend and charge that wholesale recruiting is damaging universities intellectually, because it emphasizes quantity over quality. However, a recent proposal by St. Catherine's College, Oxford, set up a preparatory, one-year program in Japan for college graduates that would be followed by postgraduate study at Oxford, suggesting Oxford is not beyond the fray. Because the recruitment of foreign students has become both lucrative and for some universities a survival mechanism, there is some concern in the U.K. – and in other countries such as the U.S., where recruitment also brings tuition to host institutions – that quality may be sacrificed because of the hard-sell techniques. "Increasingly institutions sell education and training as a commodity to overseas students," a recent study, *Responsible Recruitment*, states. "It is both good business practise and a requirement of business ethics that value for money be provided."[64] The recruitment efforts and increased scholarships have paid off. After a five-year dramatic decline in enrollments between 1979-83 (88,000 to 55,600), the overall numbers in 1987, 37,208, for the first time exceeded the all-time high in 1979 of 36,839.[65] However, several members of the scientific community are concerned that the quality of the S&Es, some 50 percent of the total, has not been maintained at pre-full-fee levels.

As a result of cutbacks in government funding and growing links between university research and industry, both British and foreign companies have increased their contributions to higher education in Britain by 100 percent in the past five years. This carries certain costs. An administrator for the Committee of Vice Chancellors reports that current plans will allow "special confidentiality" for industrial grants and contracts, which in turn would permit companies to delay publication of Ph.D. dissertations "where appropriate."[66] Informal reports from France, the U.S. and the U.K. strongly suggest that in anticipation of grantors' displeasure with foreign graduate students in an internationally

---

[63] Interview with Maureen Woodhall, Institute of Education, University of London, April 1988.

[64] Quoted by Maureen Woodhall in "Specific Measures and Programmes for Foreign Students," paper prepared for OECD meetings, November 1988.

[65] Woodhall at OECD meetings.

[66] Interview with Michael Powell, assistant to Sir Mark Richmond, chairman, Committee of Vice-Chancellors, June 15, 1988.

competitive economic environment, particularly those from Asian countries, a number of academic researchers have begun to question their traditional open-admissions policies; some in the applied sciences are limiting foreign participation.

*C. The United States*

The U.S. has a complex mix of fee arrangements for foreign students. Roughly two-thirds are supported by personal and family funds; 17.6 percent are financed by their home governments or universities; and those remaining depend on funding from U.S. sources, including scholarships, grants and assistantships.[67] Most state universities charge foreigners the same fees as out-of-state students, although 10 percent have begun to charge the full cost. In 1987 Massachusetts legislators – despite strong opposition from local academics – passed a bill requiring full-cost fees for foreigners. The argument advanced in favor of full fees stemmed from antipathy to Libya and Iran for taking U.S. hostages. However, the bill has not yet been implemented because of court appeals and the difficulty of calculating those costs. A similar bill is also being appealed in Louisiana. More recently cutbacks in Massachusetts state funds have increased the pressure to raise tuition across the board. At the graduate level most S&Es, both domestic and foreign, are currently supported by research grants or teaching and research assistantships.[68]

In one sense, full-fee tuition does not cover the real costs of a science and engineering education in any of the host countries. The average engineering Ph.D. at a public university is estimated to cost the U.S. taxpayer $100,000 to $150,000 above students' fees.[69] However, if that person later works as an engineer in the U.S., there is "a prospective return of at least 500 percent per year on that standing investment."[70] Since the majority of noncitizens earn their bachelors' degrees abroad – only 9 percent of the undergraduate population in engineering is foreign; less than 3 percent in other fields combined – and since more than 60 percent remain in the U.S. work force, they in fact represent a bargain for the U.S. economy.

The U.S. is the only country in the study that conducts large-scale government-sponsored defense research in its universities, which presents a major complication in its policies toward foreign S&Es. The Departments of Defense and

---

[67] Zikopoulos, op. cit., p. 33-34.

[68] The willingness of Chinese S&Es, for example, to work longer hours for lower pay has provoked resentment of unfair competition. A dean at the University of Arizona complains that U.S. graduate students are being discriminated against, as faculty prefer the harder-working Asian students. The Chinese will return home while the Americans, he argues, will graduate with inadequate research training, thereby weakening the country's technical force.

[69] These figures were discussed by the National Academy of Engineering Committee on Foreign Engineers, but because they could not be substantiated, they were omitted from the final report, *Foreign and Foreign-Born Engineers*, op. cit.

[70] Peter Cannon, "Foreign Engineers in U.S. Industry: An Exploratory Assessment," *Foreign and Foreign-Born Engineers*, op. cit., p. 119.

Energy sometimes officially or privately stipulate that foreign S&Es are not to be supported by their grants or contracts. In laboratories where the senior investigator has multiple funding sources, foreign students can be shifted to other grants or contracts. But where there is no such flexibility, foreign students may be turned away. Of more concern, the current cutback in government funding may lead senior researchers to choose students less apt to rile the funders. Consequently, as several faculty interviewed for the study reported, a de facto policy excluding foreign S&Es from research participation could result, even if academics do not overtly support it.

Contradictions abound. One professor who works on advanced technology with military and industrial applications accepts no foreign students, while his university actively recruits foreign S&Es as well as European and Japanese investment in its research.

Until 1989 Carnegie Mellon resisted enrolling foreign graduate students in engineering; the president recently reported the policy has been reversed.

Almost a decade ago the University of Illinois stopped accepting foreign engineering undergraduates, except for a few, approximately ten, in an undergraduate population of 5300, who were sponsored by the federal Assistance for International Development (AID) program. The policy was adopted in 1981 in response to Illinois industrialists who were lobbying for state legislation; companies claimed they had difficulty getting work permits for foreign graduates of the university and thought the university should enroll more U.S. S&Es. The school launched an intensive recruitment program, and a senior administrator in the School of Engineering reported that "Now industry can count on eighty percent of the graduates being American citizens. An active recruitment policy brought about an astounding (high) level of quality. We're dancing to a different tune from other universities."[71]

Although the majority of administrators and academics interviewed for this study in the U.S. have hastened to add that it is language barriers and immigration restrictions that have brought about attempts to control the numbers of foreign S&Es, Europeans as well as Asians have complained that foreign students are unwelcome in a number of university laboratories as the U.S. government has taken an increasingly restrictive stance toward scientific and technical communication.[72]

Concerned about the rumors of secrecy and quotas, the Institute of International Education conducted a study of U.S. engineering faculty that found the following for the 1984-85 academic year:

- 13 percent reported that some foreign students were unable to participate in research programs on grounds of national security;
- 6 percent reported that foreign students were unable to participate on grounds of

[71] Telephone interview with Dr. Howard L. Wakeland, associate dean for programs in engineering, School of Engineering, University of Illinois, February 16, 1988.

[72] Interview with Dr. Karl Roellofs, DAAD, Bonn, FRG, 1987.

economic competitiveness or proprietary restriction;
- 9 percent reported that these restrictions have been detrimental to their own research;
- 18 percent reported that funding sources had prohibited access for foreign research assistants;
- 25 percent reported that funding sources had discouraged access of foreign students.[73]

These figures merit attention and wider testing because more departments in universities are operating like engineering or applied science departments – that is, with links to industry, corporations and university-owned enterprises. The lines between basic and applied science, particularly since the development of molecular biology and biogenetic engineering are blurred, and the rules regarding secrecy are often ad hoc and unformulated. Many U.S. universities are drafting new guidelines for conflict-of-interest policies in order to gain more control over faculty members' financial dealings with industry. With increased scrutiny of laboratory practices, it is likely that discrimination against foreign S&Es – now almost impossible to document – could be more accurately monitored.

Perhaps the increasing demands of industry and government on universities to protect information is inevitable, and more research that cannot be safely designated as basic will have to be moved from university laboratories to government or industry laboratories. The intellectual costs would be considerable.

## Overview

In the industrialized countries – France, FRG, the U.K., and the U.S., which (with Canada) are the major trainers of foreign S&Es – governments and universities largely support an active effort to educate nonnational S&Es, particularly at the graduate level. Japan is beginning an energetic campaign to join the Western countries. But other voices – some university administrators and members of Congress, along with local governments, professional societies, industry, defense organizations and student groups – are raising conflicting interests and goals. Some favor more liberal and others more restrictive policies regarding which and how many foreign S&Es should be educated, for what purposes and at whose expense.

On the one hand, some side with best-selling author and sales trainer Tom Peters, who, noting the U.S. need for foreign talent, argued that foreign S&Es should receive green cards stapled to their diplomas. On the other, many of the traditional assumptions behind completely open admission of foreign S&Es are being challenged – less in print than in

[73] Barber, op. cit., p. 33-37.

seminars, workshops and informal exchanges – much more frequently than when this study began three years ago. During the course of the interviews for this study, opinions about the salience of the foreign S&E student issue ranged from "this is a non-issue" to "this is an issue about which the government (or university or industry) is very seriously concerned." There is considerably more agreement about the absence of convincing data. Both qualitative and quantitative data are lacking; both are needed. Numbers alone cannot measure the contributions of foreign S&Es to science, the economy or diplomacy. Anecdotes, no matter how telling, can be applied only with great caution. However, until the positions of all interest groups are identified, policies now in place or being formulated will lack the breadth necessary to maintain the international mobility of S&Es and to cope with the rapid dissemination of technology. With the exception of intractable xenophobia, there is a measure of reason in each point of view. What is less clear is how much weight should be given to the often mutually exclusive policy recommendations.

Within a decade all of the countries in this study are likely to be competing for S&Es, as the numbers of college-age students in industrialized countries will have declined since the post World War II baby boom by as much as 25 percent. In the U.S., by 1995, one in five present science faculty members and one in four in engineering will have reached age sixty-five.[74] In addition, current sender regions, such as Southeast and East Asia, are building up strong science and engineering departments in their own universities and will also enter the bidding for students and faculty. Since 1980 more than 20,000 U.S.-educated South Koreans have returned home, as have a significant number of Taiwanese. The promise of substantial salaries, new equipment and a large research budget at home could very well reverse the brain-drain pattern. Consequently, it is important for the vitality of science that both sender and host countries work together to acknowledge the tension between nationalism and internationalism that is threatening the education of foreign S&Es. Somewhat ironically, industrialized nations have allowed themselves to become dependent on universities for the technologies that have led to greater economic and military security. Yet the more successful academic scientists are in developing cutting-edge technology, the more their governments and national industries want to extract a toll from them in the form of greater restrictions, which may in the long run weaken the autonomy that has made them so productive.

Just raising the subject of the education of foreign S&Es is attacked in some quarters as a potentially destructive act, one which by calling attention to it could lead governments or universities to increase their control over institutional policies.[75] In a paper presented at a recent Seminar on Higher Education and the Flow of Foreign Students by the Organisation for Economic Co-operation and Development (OECD), a U.S. participant wrote:

---

[74] William G. Bowen and Julie Ann Sosa, *Prospects for Faculty in the Arts and Sciences: A Study of Factors Affecting Demand and Supply 1987-2012* (Princeton, N.J.: Princeton University Press, 1989).

[75] A reviewer of a paper on foreign S&Es submitted to *Science* by Dorothy S. Zinberg stated that publishing the negative information could lead to a tightening of existing policies.

It is not uncommon to hear an institutional or system administrator state a feeling of concern about the public reaction if 'the public' knew of the number of foreign students or the subsidies and other costs that are encumbered in support of foreign students. In fact, it is not too strong to say there has been an implicit – and at times explicit – attempt to keep foreign student issues from becoming a topic of political debate at local and state levels.

One can sympathize with this view without endorsing it. In a time of neoprotectionist trade sentiments and of legislative calls for a more self-interested and isolationist orientation to foreign relations, there is a great opportunity for political posturing and demagoguery on the foreign student issue. However, the greatest danger to the institution is not from having questions raised, as inevitably they will be. Rather, the danger is in the lack of detailed information that supporters of foreign-student participation have to justify both the benefits and to clarify the costs of such participation.[76]

As many clashing views as can be identified should be gathered in order to grasp the complexity of the issue on maintaining the balance between national needs and international participation. The balance of variables in this new mix will determine the near future of science, engineering and higher education and for the more distant future of all the various intranational relations (between government, industry, and university) and international relations (among Allied nations and among industrial, newly industrializing and third world nations).

Though perhaps seemingly innocuous, the policies underlying the education of foreign S&Es are intertwined with a broad range of complex issues – from education, economic, immigration and foreign policy to national security and economic competition. Accordingly, the policy contradictions that hinder S&Es from moving freely across international borders should be resolved to enable nations to benefit from the scientific skills needed to achieve national and global goals.

## Acknowledgements

The research for this paper was supported by the National Science Foundation. I want to thank Jennifer Bond, the senior officer who administered the grant, for her knowledge of the subject and for the direction she provided for the research. Many academics whom I interviewed contributed to the work in addition to those quoted in the body of the paper. In particular, I also want to thank some of the many people who willingly spent a great

---

[76] Quoted by Douglas M. Windham and Alan P. Wagner, "Measures and Impacts of Foreign Student Participation in the United States Higher Education: Policy, Practise and Research Perspectives," Research Institute for Higher Education Hiroshima University, OECD Conference on Foreign Students and Internationalization of Higher Education No. 8010, 1988, p. 14.

deal of time setting up appointments, providing data, and discussing the ramifications of foreign student issues – Maureen Woodhall, Lynn Williams, Eric Ash, John Ashworth, and Derek Roberts in the U.K.; Hubert Curien, Albert Lombroso, Claude Maury, and Dominique Martin-Rovet in France; Gert Fieguth, Karl Roellofs, Hans Joachim Queisser, and Manfred Stassen (U.S.) in the Federal Republic of Germany; Fumio Kodama and Shigetaka Seki in Japan; and Robert Kaplan; Erich Bloch, former director of the NSF, and my colleague, Harvey Brooks, whose erudition across a vast array of science and social science topics was brought to bear on all stages of the project.

# Appendix A

**Country of Origin of Foreign Students Within Selected Leading Host Countries**

| Leading Place of Origin | Host Country: United States[1] | France[2] | Germany, F.R[3] | United Kingdom[4] | Canada[5] |
|---|---|---|---|---|---|
| China | 29,040 | 1,488 | 1,877 | 5,600 | 2,657 |
| Taiwan | 28,760 | 3,280 | 658 | 157 | 163 |
| Malaysia | 16,170 | 154 | — | 5,600 | 1,412 |
| Iran | 8,950 | 4,414 | 8,793 | 2,023 | 249 |
| Greece | 4,360 | 2,663 | 6,208 | 2,289 | 310 |
| Morocco | 1,020 | 23,070 | 294 | 49 | 489 |
| Korea, Rep. of | 20,610 | 1,258 | 3,340 | 183 | 178 |
| Jordan | 4,940 | 317 | 963 | 591 | 62 |
| Hong Kong | 10,560 | 42 | — | 6,935 | 5,843 |
| Germany, F.R. | 6,340 | 3,660 | — | 1,496 | 402 |
| United States | — | 3,473 | 7,970 | 3,438 | 2,960 |
| India | 23,350 | 232 | 565 | — | 898 |
| Japan | 24,000 | 851 | 1,158 | 402 | 319 |
| Canada | 16,030 | 1,067 | 424 | 805 | — |
| Italy | 2,370 | 1,492 | 2,307 | 464 | 79 |
| Nigeria | 6,150 | 587 | 235 | 3,221 | 216 |
| Syria | 1,710 | 3,008 | 450 | 176 | 34 |
| Turkey | 3,010 | 763 | 9,790 | 190 | 109 |
| Lebanon | 5,130 | 4,794 | 307 | 182 | 165 |
| United Kingdom | 6,800 | 2,086 | 2,002 | — | 572 |
| Indonesia | 8,720 | 402 | 2,131 | — | 385 |
| Algeria | 580 | 10,062 | 300 | 562 | 235 |
| France | 4,880 | — | 2,650 | — | 1,073 |
| Egypt | 1,850 | 715 | 761 | 622 | 153 |
| Sudan | 570 | 118 | — | — | 21 |
| Tunisia | 720 | 7,131 | 485 | — | 370 |
| Cyprus | 1,710 | — | 658 | — | 33 |
| Spain | 3,050 | 2,768 | 1,804 | 234 | 81 |
| Singapore | 4,460 | 33 | — | 1,271 | 1,136 |
| Pakistan | 7,050 | 51 | 199 | — | 120 |
| Cameroon | 1,410 | 4,796 | 311 | 193 | 313 |
| Saudi Arabia | 4,970 | 62 | — | — | 176 |
| Mexico | 5,780 | 580 | 256 | — | 231 |
| Venezuela | 3,040 | 291 | 134 | 109 | 86 |
| Iraq | 800 | 432 | 386 | 1,568 | 53 |
| Brazil | 3,300 | 1,119 | 633 | — | 203 |
| **Total, Leading Senders** | **272,190** | **87,259** | **58,049** | **38,360** | **21,786** |
| **% of All Foreign Students** | **74.3** | **70.4** | **71.6** | **68.4** | **74.0** |
| **Total, All Foreign Students** | **366,354** | **123,978** | **81,090** | **56,121** | **29,437** |

[1] *Source:* Open Doors 1988.
[2] *Source: Ministere de lEducation Nationale, Tableaux Statistiques des Etudiants Inscrites dans les Etablissments Universaires, 1987-88.*
[3] *Source: Federal Ministry of Education and Science Germany F.R., Basic and Structural Data 1988/89.*
[4] *Source: Statistics of students from abroad in the United Kingdom 1984/85, The British Council.*
[5] *Source: Canadian Bureau for International Education; figures are for 1988/89.*

Source: Marianthi Zikopoulos, ed., *Open Doors 1988-1989*, (New York: Institute of International Education, 1989).

## Appendix B

**Foreign Student Enrollment in Leading Host Countries, Selected Years, 1970-1986**

| Host Country | 1970 | 1975 | 1980 | 1984 | 1985 | 1986 |
|---|---|---|---|---|---|---|
| United States[1] | 144,708 | 179,350 | 311,882 | 342,113 | 343,777 | 349,610 |
| France | 34,500 | 93,750 | 110,763 | 133,848 | 131,979 | 126,762 |
| Germany, F.R. | 27,769 | 53,560 | 61,841 | 76,918 | 79,354 | — |
| U.S.S.R. | 27,918 | 43,287 | 62,942[2] | 42,267[4] | — | — |
| United Kingdom | 24,606 | 49,032 | 56,003 | — | — | — |
| Canada | 22,263 | 22,700 | 28,443 | 32,625 | 29,496 | 27,210 |
| Lebanon | 22,184 | — | 31,028 | 25,515 | — | — |
| Italy | 14,357 | 18,921 | 29,447 | 27,548 | — | — |
| Egypt | 13,387 | — | 21,751 | — | 12,235 | — |
| Saudi Arabia | 1,404 | 4,026 | 14,298 | 17,283 | 17,607 | — |
| Switzerland | 9,469 | 10,113 | 14,716 | 16,830 | 17,396 | 13,579 |
| Romania | 1,766 | 4,971 | 15,888 | 13,068 | 10,774 | — |
| India | 7,804 | 8,880 | 14,710[3] | — | — | — |
| Austria | 8,573 | 10,320 | — | 14,858 | 15,388 | 15,740 |
| Sweden | — | — | 13,182 | — | — | — |
| Belgium | 8,611 | 9,748 | 12,875 | 21,188 | 24,761 | 20,095 |
| Spain | 10,575 | 8,909 | 10,997 | — | — | — |
| Brazil | — | — | 12,800 | — | — | — |
| Australia | 7,525 | 8,356 | 8,777 | 12,028 | 16,075 | — |
| Argentina | — | — | 7,882[3] | 10,049 | — | — |
| Japan | 4,447 | 5,541 | 6,543 | 10,692 | 12,442 | 14,960 |
| Vatican City | 8,128 | 5,740 | 9,104 | 9,656 | 9,775 | — |
| German D.R. | — | 5,386 | 7,106 | 9,143 | 9,231 | 9,613 |
| Yugoslavia | — | 2,358 | 4,426 | 7,982 | — | 7,426 |
| Greece | 5,796 | 1,049 | 7,673 | — | — | — |
| Turkey | 6,125 | 5,907 | — | 6,732 | 7,021 | 6,923 |

* *Source:* Unesco Statistical Yearbook 1988, *Table 3.11 "Education at the third level: number of foreign students enrolled", pp. III 355-359.*
[1]*Source:* Open Doors, 1986/87.
[2]*Figures are for 1978.*
[3]*Figures are for 1979.*
[4]*Figure is for 1983.*

Source: Marianthi Zikopoulos, ed., *Open Doors 1988-1989,* (New York: Institute of International Education, 1989).

# Appendix C

**Distribution of Foreign Students Among Host Countries by Leading Place of Origin[1]**

| Host Country[2] | Country of Origin | | | | | | | | | |
|---|---|---|---|---|---|---|---|---|---|---|
| | China[3] | Malaysia | Iran | Greece | Morocco | Korea, Rep. of | Jordan | Hong Kong | Germany, F.R. | United States |
| United States (1986) | 75.5 | 49.0 | 29.1 | 10.9 | 2.5 | 67.7 | 20.4 | 40.0 | 19.4 | — |
| France (1986) | 2.4 | 0.4 | 13.0 | 8.5 | 79.9 | 4.8 | 1.2 | 0.2 | 16.0 | 16.5 |
| Germany F.R. (1985) | 2.1 | 0.1 | 21.2 | 19.5 | 0.7 | 10.3 | 4.0 | 0.0 | 0 | 19.6 |
| United Kingdom (1984) | 0.7 | 11.4 | 4.7 | 6.2 | 0.2 | 0.8 | 1.9 | 23.7 | 6.6 | 16.4 |
| Italy (1983) | 0.1 | 0.0 | 9.5 | 35.7 | 0.0 | 0.0 | 5.8 | 0.0 | 7.5 | 4.7 |
| Canada (1986) | 2.8 | 4.1 | 0.6 | 0.9 | 2.3 | 0.6 | 0.2 | 27.7 | 1.4 | 12.0 |
| Lebanon (1984) | 0.0 | 0.0 | 0.0 | 0.0 | 0.0 | 0.0 | 24.0 | 0.0 | 0.0 | 0.0 |
| Belgium (1986) | 0.4 | 0.3 | 0.8 | 2.2 | 8.3 | 0.1 | 0.0 | 0.0 | 2.6 | 1.3 |
| Saudi Arabia (1986) | 0.1 | 0.2 | 0.1 | 0.0 | 0.4 | 0.0 | 7.0 | 0.0 | 0.0 | 0.2 |
| Australia (1985) | 0.3 | 19.6 | 0.0 | 0.0 | 0.0 | 0.4 | 0.0 | 6.9 | 0.4 | 1.4 |
| Austria (1986) | 0.5 | 0.0 | 3.8 | 1.6 | 0.0 | 1.0 | 0.2 | 0.0 | 19.3 | 1.9 |
| Japan (1986) | 13.1 | 1.7 | 0.1 | 0.0 | 0.0 | 12.1 | 0.0 | 0.7 | 0.3 | 4.3 |
| Switzerland (1986) | 0.2 | 0.0 | 0.8 | 1.2 | 0.5 | 0.2 | 0.1 | 0.0 | 13.2 | 2.5 |
| Syria (1985) | 0.0 | 0.0 | 0.5 | 0.0 | 0.1 | 0.0 | 12.6 | 0.0 | 0.0 | 0.0 |
| Egypt (1985) | 0.0 | 0.0 | 0.1 | 0.1 | 0.0 | 0.0 | 2.4 | 0.0 | 0.0 | 0.0 |
| Spain (1980) | 0.0 | 0.0 | 1.7 | 0.3 | 0.9 | 0.0 | 1.2 | 0.0 | 0.9 | 3.0 |
| Sweden (1984) | 0.0 | 0.0 | 2.5 | 1.3 | 0.1 | 0.1 | 0.1 | 0.0 | 1.6 | 1.5 |
| Vatican City (1985) | 0.0 | 0.0 | 0.0 | 0.0 | 0.0 | 0.4 | 0.0 | 0.0 | 1.1 | 3.4 |
| Germany, D.R. (1986) | 0.0 | 0.0 | 0.0 | 0.0 | 0.0 | 0.0 | 0.0 | 0.0 | 0.0 | 0.0 |
| Yugoslavia (1986) | 0.0 | 0.0 | 0.7 | 9.3 | 0.0 | 0.0 | 3.6 | 0.0 | 0.2 | 0.1 |
| % share, 20 Leading Countries | 98.4 | 87.0 | 89.3 | 97.7 | 96.1 | 98.5 | 84.9 | 99.3 | 90.5 | 88.8 |
| Total, 50 Leading Countries | 100.0 | 100.0 | 100.0 | 100.0 | 100.0 | 100.0 | 100.0 | 100.0 | 100.0 | 100.0 |

| Host Country[2] | Country of Origin | | | | | | | | | |
|---|---|---|---|---|---|---|---|---|---|---|
| | India | Japan | Canada | Italy | Nigeria | Syria | Turkey | Lebanon | United Kingdom | Algeria |
| United States (1986) | 79.4 | 74.2 | 80.5 | 10.4 | 62.4 | 10.2 | 13.6 | 36.7 | 37.4 | 4.8 |
| France (1986) | 1.2 | 4.3 | 6.1 | 8.4 | 3.9 | 18.8 | 4.7 | 33.2 | 13.6 | 77.2 |
| Germany F.R. (1985) | 2.9 | 6.3 | 2.5 | 13.3 | 1.7 | 2.8 | 62.1 | 2.0 | 13.3 | 2.7 |
| United Kingdom (1984) | 4.1 | 2.1 | 4.7 | 2.6 | 17.3 | 0.9 | 1.0 | 1.0 | 0.0 | 4.3 |
| Italy (1983) | 0.4 | 0.0 | 0.5 | 0.0 | 3.1 | 1.5 | 0.8 | 4.4 | 1.2 | 0.0 |
| Canada (1986) | 4.1 | 1.1 | 0.0 | 0.5 | 1.3 | 0.2 | 0.5 | 0.9 | 8.4 | 1.5 |
| Lebanon (1984) | 0.0 | 0.0 | 0.0 | 0.0 | 0.0 | 38.1 | 0.0 | 0.0 | 0.0 | 0.0 |
| Belgium (1986) | 0.4 | 0.1 | 0.4 | 14.3 | 0.3 | 0.6 | 1.7 | 3.2 | 1.6 | 3.0 |
| Saudi Arabia (1986) | 1.4 | 0.0 | 0.0 | 0.0 | 0.8 | 5.4 | 1.3 | 0.7 | 0.1 | 0.3 |
| Australia (1985) | 1.0 | 1.1 | 0.7 | 0.1 | 0.3 | 0.0 | 0.0 | 0.0 | 1.8 | 0.0 |
| Austria (1986) | 0.4 | 1.1 | 0.3 | 17.6 | 0.6 | 0.5 | 5.0 | 0.2 | 0.9 | 0.2 |
| Japan (1986) | 0.5 | 0.0 | 0.4 | 0.2 | 0.1 | 0.0 | 0.1 | 0.0 | 0.7 | 0.0 |
| Switzerland (1986) | 0.3 | 0.3 | 0.9 | 10.3 | 0.1 | 0.3 | 1.2 | 0.7 | 2.2 | 0.7 |
| Syria (1985) | 0.0 | 0.0 | 0.0 | 0.0 | 0.0 | 0.0 | 0.7 | 7.9 | 0.0 | 1.3 |
| Egypt (1985) | 0.0 | 0.0 | 0.0 | 0.1 | 0.0 | 1.0 | 0.0 | 0.6 | 0.0 | 0.1 |
| Spain (1980) | 0.1 | 0.1 | 0.0 | 0.8 | 0.1 | 1.9 | 0.0 | 1.0 | 0.6 | 0.0 |
| Sweden (1984) | 0.2 | 0.2 | 0.2 | 0.5 | 0.2 | 0.1 | 0.9 | 0.0 | 1.7 | 0.2 |
| Vatican City (1985) | 2.7 | 0.2 | 0.7 | 18.4 | 3.0 | 0.1 | 0.0 | 0.7 | 1.2 | 0.0 |
| Germany, D.R. (1986) | 0.0 | 0.0 | 0.0 | 0.0 | 0.0 | 0.0 | 0.0 | 0.0 | 0.0 | 0.0 |
| Yugoslavia (1986) | 0.0 | 0.0 | 0.1 | 0.0 | 0.2 | 5.9 | 0.5 | 0.8 | 0.0 | 1.2 |
| % share, 20 Leading Countries | 99.0 | 91.4 | 98.1 | 97.3 | 95.4 | 88.2 | 94.3 | 94.0 | 84.7 | 97.5 |
| Total, 50 Leading Countries | 100.0 | 100.0 | 100.0 | 100.0 | 100.0 | 100.0 | 100.0 | 100.0 | 100.0 | 100.0 |

*Source: Unesco* Statistical Yearbook 1988, *Table 3.12 "Education at the third level: foreign students by country of origin in 50 selected countries", pp. III 360-385.*

[1] *Percentages are based on the student total from each country of origin. Blanks indicate less than 0.1%.*

[2] *Dates in parentheses represent the year on which percentages are based. Host countries that reported large numbers of foreign students, but did not report the necessary country-of-origin data are omitted from this table: Soviet Union, Romania, Sweden, and Brazil.*

[3] *Unesco data on China include figures for Taiwan.*

Source: Marianthi Zikopoulos, ed., *Open Doors 1988-1989*, (New York: Institute of International Education, 1989).

Appendix D

**Foreign Students by Field of Study, 1988/89**

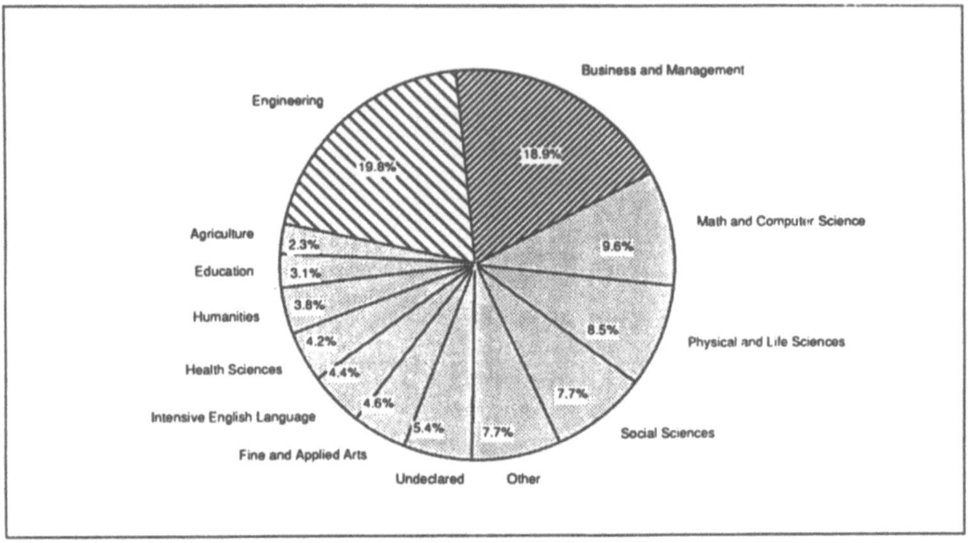


Source: Marianthi Zikopoulos, ed., *Open Doors 1988-1989*, (New York: Institute of International Education, 1989).

# Appendix E

## SUMMARY FOREIGN STUDENT ENROLLMENTS *

| Receiving Country | Number of Foreign Students Enrolled in Post Secondary Education | Percent total Post-Secondary Enrollment | Major Countries/ Regions of Origin | Graduate Enrollments | Leading Enrollments by Academic Subject |
|---|---|---|---|---|---|
| Great Britain | 64,400 (30% female) | 10.7% | Hong Kong, USA, Greece, Malaysia, Singapore; developing countries falling | 36.6% of all graduate enrollments; 52.4% of all graduate enrollments in science and engineering | 40% science and engineering; 25% administration and management |
| France | 124,000 (33% female) | 12.5% | 57% Africa; 83% developing | 33% of foreign students enrolled in 3rd cycle compared with 14% of French students | 36% literary studies; 23% science |
| Germany | 77,445 | 5.5% | Iran, Turkey, Greece, Indonesia; enrollments from EEC countries rising | N.A. | 27% engineering; 14% science and mathematics |
| Japan | 25,643 (university only) (30% female) | 1.2% | 88.9% Asia | 36% of foreign students enrolled at graduate levels; heavy concentrations in several major universities | 20.3% engineering |
| Australia | 17,655 (36% female) | 4.5% | 88% Asia; 6% Oceania (50% Malaysia) | 25% of foreign students enrolled at graduate level | 50% undergraduates business administration, economics, engineering |
| Canada | 37,100 (enrollments declining) (35% female) | 4.8% | Hong Kong, USA, PRC, Malaysia, Singapore | 10% of foreign students enrolled at graduate level | 20% commerce, management, business administration, economics; 11% computer science |

* Based on most recent available data for each country. Not all data bases are completely comparable.

Source: Alice Chandler, *Obligation or Opportunity: Foreign Student Policy in Six Major Receiving Countries*, IEE Research Report No. 18 (1989).

# SCIENCE AND ENGINEERING: HUMAN RESOURCE NEEDS IN THE NEXT THREE DECADES

ALAN FECHTER
*National Research Council*[1]

## Abstract

The changing composition of the workforce in the United States raises fears that the country will soon suffer from a shortage of scientists and engineers. The author analyzes survival rates of college students and doctoral students to show that nondemographic trends must be investigated before we can understand the impact of the changing population on the production of scientists and engineers. He then discusses two other models of the projected supply and demand for engineers.

## Introduction

This paper will examine the demographic realities confronting the United States and the issues they raise about the future adequacy of our supply of scientists and engineers (S&Es). The analysis will be restricted to natural scientists and engineers (NSE), although much of the discussion will also apply to behavioral and social sciences. Where possible, the analysis will address scientists and engineers separately and treat those with bachelor's degrees separately from those with doctorates, since the markets and issues for these groups are quite different.

These facts are familiar enough to be conventional wisdom:

– From 1957 to 1973 the United States experienced a 28 percent decline in births. This has resulted in a significant reduction in the size of the pool from which we will draw our future supply of talent.[2]

– Minority groups – particularly those who are underrepresented in science and engineering fields – will be an increasingly prominent portion of this pool, both because of recent immigration patterns and because these groups' birth rates are higher.

---

[1] The views expressed in this paper are those of the author; they do not necessarily reflect the views of the National Research Council or its parent entities, the National Academy of Sciences and Engineering or the Institute of Medicine.

[2] National Center for Education Statistics, *Projections of Education Statistics to 1997-98*. CS 88-607, p. 117. See also p. 12.



*D. S. Zinberg (ed.), The Changing University*, 89–99.

– Women – another group that has traditionally been underrepresented in these fields – have become a larger fraction of the labor force and of the undergraduate pool from which science and engineering talent is drawn.
– The median age of scientists and engineers is increasing, suggesting that more of them will have to be replaced as they leave the work force.

A number of important conclusions flow from these realities. Given the declining pool and the growing importance of women and minorities as a share of that pool, future recruitment of S&Es will be more difficult for the same level of demand. However, demand can only be expected to increase, given the number of older workers who will comprise the S&E work force and the projected increases in requirements for these skills.

The policy question that emerges from these conclusions is: Will we be able to meet our future needs for S&E talent? A rational answer requires factual information on a number of topics, including the following:

– How close is the link between changes in the labor pool and changes in new S&E degrees?
– Given the nature of this linkage, will there be enough new S&E entrants to meet projected needs?

This paper addresses these two questions. It begins with an examination of recent trends in degree production and the determinants of these trends. It then turns to an examination of the future and assesses both short-term and long-term projections.

## Demography and the Production of New Supply

A considerable amount of attention has been devoted to the link between the size of the pool from which new degree recipients are drawn and the number of new degree recipients. The typical study has defined the cohort of 22-year olds as the pool from which undergraduate degree recipients are drawn and the cohort of 30-year olds as the pool from which doctorates are drawn.[3] These degree recipients are also drawn in large numbers from adjacent cohorts. In 1986, for example, the modal age of high school seniors was 17.[4] Thus, definitions based on a single-year age cohort can only serve as rough approximations of the pool.

Nevertheless, statistics reveal a remarkable stability in the number of undergraduate

---

[3] See, for example, U.S. Congress, Office of Technology Assessment, *Educating Scientists and Engineers: Grade School to Grad School*, OTA-SET-377 (Washington, D.C.: U.S. Government Printing Office, 1988): 8-10; National Science Foundation, Division of Policy Research and Analysis, *The Science and Engineering Pipeline*. PRA Project 87-2, April, 1987, p. 4-5,13-17.

[4] U.S. Bureau of the Census, *Current Population Reports*, Series P-20, "School Enrollments," No. 429.

degrees produced in natural science and engineering fields per 100 22-year olds in the population. This ratio averaged slightly more than four for the 20 year period from 1959 to 1979, with relatively little annual variation from this mean. It rose by roughly one-fourth between 1980 and 1986 to slightly more than five, fueled largely by the surge in the number of computer science degrees granted.

In striking contrast, the link between doctorates in natural science and engineering and the size of the pool is much more tenuous. The number of U.S. citizens earning doctorates in these fields declined by about one-third from 1971 to 1978, a period when the 30-year old population was rising.

In both instances, using the population cohort as a basis for predicting degree production can be a poor way to understand the dynamics that underlie the process. A degree is the outcome of a journey through the educational process in which success almost invariably requires comparable success at attaining prior levels of education. Undergraduate training commences after completion of high school; graduate training typically begins after receipt of an undergraduate degree.

Given the cumulative nature of this process, an alternative mechanism would take into account the number of students who complete the earlier levels of schooling – for example, bachelor's degree recipients would be measured in relation to high school graduates four years earlier, and doctorate recipients in relation to bachelor's degree recipients six to eight years earlier, depending on the field. This ratio would more closely approximate the longitudinal nature of the process. It would also make possible current short-term projections of degree production from observations of the pool.

Recognition of this longitudinal, cumulative training process enables one to specify degree production as a function of prior attainment rates and the relevant pool, appropriately lagged. For example, the number of baccalaureates granted in a given year could be expressed as the product of the pool from which high school graduates are drawn, lagged four years, and the current degree attainment rate of this pool. The latter rate will be the product of high school graduation rates, continuation rates of these graduates to college, and degree completion rates of those who continue.

Expressing degree production in these terms allows us to better understand its variation. The annual number of bachelor's degrees in natural sciences and engineering increased 38 percent between 1973 and 1986, rising from 154,500 to 213,900 (Chart 1). The annual average growth rate of 2.5 percent for this period hides a significant amount of variation over time and by field, however. The annual growth rate was considerably lower in the 1970s, for example – 1.9 percent. It was considerably higher in the 1960s, 5.2 percent. In addition, the overall growth was the result of dramatic increases in degree production in engineering and computer sciences, which mask a distinct long-term downward trend for the natural sciences.

What were the major forces driving this growth? The pool of high school graduates grew by only four percent over this period (Table 1). This means that about one-tenth of the increase in degree production was attributable to growth in the size of the pool. To assess the combined effects of nondemographic factors – i.e., continuation to college and completion of degrees in these fields – I estimated a baccalaureate "survival" rate, the number of bachelor's degrees awarded in natural science and engineering fields per hundred high school graduates four years earlier. This rate rose by more than 30 percent between 1973 and 1986.

*Table 1.* Determinants of Changes in NSE Bachelor's Degree Production, 1973-1986

| | Number of degrees (in thousands) | Number of high school graduates* (in thousands) | Survival rate** |
|---|---|---|---|
| 1973 | 157.1 | 2888 | 5.44 |
| 1986 | 213.9 | 3001 | 7.13 |
| *Percent change,* 1973-1986 | +36.1 | +3.9 | +31.1 |

*Source*: Appendix Table 1
*Notes*:
*Lagged 4 years
**[Column (1) divided by Column (2)] x 100

From these findings, I conclude that factors other than the size of the pool play an important role in determining the number of bachelor's degrees granted in these fields. This analysis implies that projections of future S&E bachelor's degrees based solely on demographic factors are fundamentally flawed and must be treated with great caution.

Similarly, one can better understand variation in the annual number of natural science and engineering doctorates by examining survival rates of bachelor's degree recipients of earlier periods. I have confined my analysis to natural sciences and engineering doctorates received by U.S. citizens. The number of doctorates in these fields remained remarkably stable between 1977 and 1987, ranging between 8,500 and 9,000. It rose from 8,700 to 9,400 between 1986 and 1988. As noted earlier, the number of bachelor's degrees in the natural sciences rose 45 percent between 1970 and 1986. Thus, the trends in doctoral degrees over this period must be reflecting declines in survival rates of college graduates.

I crudely approximated these rates by the number of doctorates per 100 S&E bachelor's degrees awarded seven years earlier. The stability in doctorate degrees between 1977 and 1987 reflects a growing pool of S&E bachelor's degree recipients offsetting a declining doctoral survival rate (Table 2). The recent rise in the number of doctorates granted in these fields was largely the result of the reversal of the decline in survival rates between 1987 and 1988, although it also reflects the growth in bachelor's degrees that began in the 1980s.

A major finding from this empirical assessment is that when the analysis conforms more closely to the education process that produces these degrees, nondemographic factors become an important determinant of change for both bachelor's degrees and doctorates. Rational policy formulation and accurate projection of degree production thus requires a better understanding of these factors.

*Table 2.* Determinants of Change in NSE Doctorate Production, U.S. Citizens, 1975-1988

| | Number of doctorate degrees (in thousands) | Number of NSE bachelor's degrees* (in thousands) | Survival rate** |
|---|---|---|---|
| 1975 | 9.3 | 124.5 | 7.5 |
| 1985 | 8.8 | 167.8 | 5.3 |
| 1986 | 8.7 | 172.8 | 5.0 |
| 1987 | 8.9 | 177.1 | 5.0 |
| 1988 | 9.4 | 182.9 | 5.2 |
| *Percent change* | | | |
| 1975-85 | -5.4 | +34.8 | -29.4 |
| 1985-88 | +6.8 | + 9.0 | - 1.9 |
| 1975-88 | +1.0 | +46.9 | -30.7 |

*Source*: Appendix Table 2.
*Notes*:
*Lagged 7 years.
**[Column (1) divided by Column (2)] x 100

## Future Trends

Given these patterns, what can we say about the future? Survey data on career plans of college freshmen can provide some clues about the immediate future. The fraction who indicate interest in natural sciences or engineering has been declining (Chart 3). Inferences from these findings must be tempered, however, because of seemingly contradictory attitudes of freshmen concerning interest in acquiring doctorates, which has been increasing in the 1980s (Chart 4).

The waning interest among freshmen in S&E majors suggests that the recent upward trend in survival rates for S&E bachelor's degree recipients will soon be reversed – particularly in engineering and computer sciences. Given the decline in the lagged pool of high school graduates that will occur between now and 1992, this means that degree production will almost certainly decline in the immediate future. The rising proportion of freshmen interested in doctorates, on the other hand, makes near-term projections more problematic. That increased interest may continue the upward movement in survival rates observed for 1986-1988 (see Appendix Table 2).

What can be expected in the longer term? The Bureau of the Census projects a continuing decline in the 18-year old population from 1988 through 1992 (Table 3), then a reasonably steady increase between 1993 and 2005 – at which time the number will have returned to 1988 levels – and then a gentle decline from 2005 to 2015. At that time the 18-year old population will be roughly five percent below the 1988 value. Given these trends, any future growth in degree production will likely come from increases in survival rates.

*Table 3.* Projections of the Number of 18-Year Olds in the U.S. Population,* 1985-2020 (in thousands)

| Year | Number | 1985 = 100 |
|---|---|---|
| 1985 | 3,658 | 100.0 |
| 1990 | 3,431 | 93.8 |
| 1995 | 3,332 | 91.1 |
| 2000 | 3,751 | 102.5 |
| 2005 | 3,943 | 107.8 |
| 2010 | 3,877 | 106.0 |
| 2015 | 3,655 | 99.9 |
| 2020 | 3,584 | 98.0 |

*Source*: U.S. Bureau of Census, *Current Population Reports*, Series P-25, No. 952, Projections of the Population of the United States by Age, Sex, and Race, 1983 to 2020, U.S. Government Printing Office, Washington, D.C., 1984, Table 6.

*Note*:
*Numbers are as of July 1 each year and include armed forces overseas.

Because of the paucity of research on the determinants of these survival rates, projections of S&E bachelor's degrees in these fields are uncertain at best and should be treated with considerable caution. The same goes for projections of doctorates, only the range of uncertainty will be even greater, since these projections are generated in part from the uncertain estimates of future bachelor's degrees.

## Will There Be Enough?

Assessing the future need for scientists and engineers is a formidable task. It is difficult enough to project these needs to the end of the century, let alone attempting to gauge what they will be 30 years from now. First, one must define need. One person's need may be another's luxury. Moreover, need can reflect narrow occupational skills or broader training requirements. This is an important distinction, since many who are now working in S&E positions do not have degrees in these fields, and many with S&E degrees are working in non-S&E jobs. Finally, we are all painfully aware of the limitations of any activity that attempts to divine the future. One must therefore offer the results of such efforts with a generous dose of humility.

Despite these problems in framing the discussion, efforts have been made to project both the future needs for and the future supply of natural scientists and engineers. The work of the National Science Foundation (NSF) has featured prominently in these discussions. At the bachelor's level, the NSF model projects degree production based largely on census projections of the population of 22-year olds.[5] The model uses 1985-1986 bachelor's degree production – the historical peak of this series – as a proxy for future needs. The shortfall for the year 2006 is estimated to be a dramatic 675,000 degrees, more than three times the peak annual production.

NSF uses a similar, though somewhat different model to estimate projected shortfalls of doctorates in these fields. It uses bachelor's degrees in these fields lagged seven years and continuation rates that are roughly comparable in concept to the survival rates employed in this paper.[6] It also assumes that 4,500 noncitizens receiving S&E doctorates from U.S. universities will remain in this country as immigrants. For doctorates, NSF defines needs in terms of job openings. Separate estimates are generated for job openings arising from deaths or retirement and job openings generated by employment growth.

Although NSF is careful to state that shortfalls are not necessarily shortages, the findings from their analyses have been interpreted as evidence of potentially serious

---

[5] See, for example, National Science Foundation, Division of Policy Research and Analysis, *Future Scarcities of Scientists and Engineers: Problems and Solutions*, working draft, Summer 1990. It also assumes that the number of degrees produced in natural science and engineering fields per 100 22-year olds will fall from its 1986 peak of 5.2 to 4.4 in 1993, after which it will rise again to 5.0 by 1997. The projected decline in this ratio is based on data describing freshman plans to major in these fields.

[6] NSF reports use a "weighted continuation rate." Unfortunately, this term is not defined in the draft manuscript.

supply inadequacies.[7]

Bowen and Sosa undertook a similar effort to assess the adequacy of the future supply of faculty.[8] Confining themselves to doctorates in fields of arts and sciences – they do not treat the professions and, consequently, they exclude engineering from their analysis – they develop an admirably explicit set of equations and variables for projecting faculty supply and demand. An important component of supply is, of course, new doctorates in these fields. They project over five-year intervals beginning with the period 1987 to 1992 and ending with the period 2012 to 2017. To their credit, they use a large number of alternative scenarios to emphasize the uncertainty associated with the results of their model. They project a relatively stable academic supply of new doctorates, ranging between 31,000 and 33,000 in each five-year period, with slight declines occurring after the 1987-to-1992 and the 1992-to-1997 periods.[9] Based on their estimates of future supply and demand, they project an excess supply from 1987 to 1997 and an excess demand afterwards (Chart 5).

These models share a characteristic that biases them in the direction of overstating supply-demand imbalances; they do not include any market feedback mechanisms. In simple terms, this means that the models assume that supply does not respond to exogenous shifts in demand and vice versa. The assumption that feedback does not occur thus makes projections generated by such models worst-case scenarios, given their parameters, particularly in the case of long-range projections. Bowen and Sosa, as well as NSF, recognize this aspect of their models and include in their studies an extensive discussion of adjustment mechanisms.

Assuming the existence of such mechanisms transforms the issue at hand. The question is no longer whether there will be enough scientists and engineers. Instead, it relates to the types of adjustments that can be expected to occur in the absence of policy intervention and the speed with which these adjustments can be expected to eliminate market imbalance.

For example, to reduce excess demand for faculty, academic administrators may allow student-faculty ratios to rise. This type of adjustment will have an immediate impact, but policymakers may wish to ask whether such increases will be socially desirable. The

---

[7] See, for example, Richard C. Atkinson, "Supply and Demand for Science and Engineering Ph.D.s: A National Crisis in the Making." Remarks to the Regents of the University of California, February 16, 1989, and "Bush's Science Adviser Gains Visibility," *Washington Post*, December 26, 1989, p. A21.

[8] William G. Bowen and Julie Ann Sosa, *Prospects for Faculty in the Arts and Sciences* (Princeton, New Jersey: Princeton University Press, 1989).

[9] Ibid., p. 102-109, 118-120. Their projections assume that production of new doctoral graduates who are U.S. citizens or noncitizens who have permanent-resident status will remain stable, that production of doctoral graduates who are noncitizens with temporary-visa status will rise slightly between the five-year periods 1987-1992 and 1992-1997, and that the relative increase in employment opportunities for new doctorates in industry will continue, but at a slower pace than experienced prior to 1987.

answer is not immediately evident. Increases in the ratio can alleviate pressure on tuition by reducing the average cost of the teaching input. But an increased student-faculty ratio may also affect the quality of the product. Larger classes are not generally perceived to be a move in the direction of higher quality education, although research does not strongly support this case.

Employers can also change their standards for hiring. For example, during periods of stringency, academic employers can hire candidates who have not completed their doctoral requirements. We saw this happen in the 1960s, the period of dramatic growth in the academic sector. During periods of plentiful supply, employers can limit their hiring to candidates who have completed their doctoral work and who have received their degrees from one of the top-ranked universities. We saw this phenomenon at work during the 1970s, when the market for academics was relatively stagnant. Such variation in standards can have significant implications for the quality and performance of new academic hires.

A similar argument can be made for the use of increased wages for equilibrating supply and demand – the "free market" solution. Such increases can be expected to reduce the supply-demand imbalance, for example, by providing incentives to students to acquire degrees in these fields, thereby increasing supply. Given the extensive amount of time required to complete training, however, this impact can be expected to be felt with a considerable lag.

Given the likelihood that such mechanisms will be triggered by any future disparity between supply growth and demand growth, sensible policy decisions require some knowledge about the choice of mechanisms and the impact these choices may have on the supply and demand for scientists and engineers, in both quantitative and qualitative terms. The literature on the determinants of these choices and their impacts is not extensive, and the small amount that exists is not very current. Clearly these issues merit further inquiry.

**Appendix**

*Table 1.* Number of High School Graduates and NSE Bachelors' Degrees Produced, 1973-1988

| Year | High School Graduates (in thousands)* | NSE Bachelor's Degree Number (in thousands) | Survival Rate (in percent)** |
|---|---|---|---|
| 1973 | 2888 | 154.8 | 5.36% |
| 1974 | 2896 | 159.6 | 5.51% |
| 1975 | 2937 | 157.1 | 5.35% |
| 1976 | 3001 | 159.6 | 5.32% |
| 1977 | 3036 | 163.5 | 5.39% |
| 1978 | 3074 | 167.0 | 5.43% |
| 1979 | 3133 | 172.8 | 5.52% |
| 1980 | 3148 | 177.1 | 5.63% |
| 1981 | 3154 | 182.9 | 5.80% |
| 1982 | 3127 | 187.2 | 5.99% |
| 1983 | 3101 | 196.9 | 6.35% |
| 1984 | 3043 | 205.6 | 6.76% |
| 1985 | 3020 | 213.9 | 7.08% |
| 1986 | 3001 | 213.9 | 7.13% |
| 1987 | 2888 | 205.8 | 7.13% |
| 1988 | 2767 | 193.5 | 6.99% |

*Notes*:
* Lagged 4 years
** [Column (2) divided by Column (1)] x 100

**Appendix**

*Table 2.* Number of NSE Doctorates Produced, NSE Doctorate Survival Rates, and NSE Bachelor Degrees Produced, 1975-1993

| | NSE Bachelor's* Degrees (in thousands) | Survival Rate (%)** | NSE Doctorates (in thousands) |
|---|---|---|---|
| 1975 | 124.5 | 7.47 | 9.3 |
| 1976 | 140.0 | 6.41 | 9.0 |
| 1977 | 147.5 | 5.90 | 8.7 |
| 1978 | 145.7 | 5.83 | 8.5 |
| 1979 | 147.5 | 5.95 | 8.8 |
| 1980 | 154.8 | 5.65 | 8.7 |
| 1981 | 159.6 | 5.50 | 8.8 |
| 1982 | 157.1 | 5.66 | 8.9 |
| 1983 | 159.6 | 5.47 | 8.7 |
| 1984 | 163.5 | 5.46 | 8.9 |
| 1985 | 167.0 | 5.26 | 8.8 |
| 1986 | 172.8 | 5.06 | 8.7 |
| 1987 | 177.1 | 5.02 | 8.9 |
| 1988 | 182.9 | 5.14 | 9.4 |
| 1989 | 187.2 | 5.12 | 9.6 |
| 1990 | 196.9 | | |
| 1991 | 205.6 | | |
| 1992 | 213.9 | | |
| 1993 | 213.9 | | |

*Sources*: NSE Doctorates: National Science Foundation, *Science and Engineering Doctorates: 1960-1988*, NSF 89-320, Table 2.

NSE Bachelor's Degrees: National Science Foundation, Division of Policy Research and Analyses, unpublished tabulations.

*Notes*:
* Lagged 7 years
** [Column (3) divided by Column (1)] x 100

# STRUCTURAL CHANGES IN THE JAPANESE SUPPLY/EMPLOYMENT SYSTEMS OF ENGINEERS: ARE WE LOSING OR GAINING?

FUMIO KODAMA and CHIAKI NISHIGATA[1]
*National Institute of Science and Technology Policy*
*Science and Technology Agency*
*Nagata-Cho 1-11-39, Chiyoda-ku*
*100 Tokyo, Japan*

## Abstract

Major changes are occurring in the supply and employment patterns of engineers in Japan. The authors analyze a unique Japanese system of producing doctoral engineers through dissertations instead of coursework; the economic implications of the growing number of engineers employed by the finance/insurance sector; the rigidity in restructuring universities despite the emergence of new disciplines such as information science; and the utilization, assignments, and promotion potential of engineers in the automotive manufacturing industry compared to their counterparts in the United States.

## Introduction

Major changes are occurring in the supply and employment patterns of scientists and engineers in Japan. In this paper, we first compare the Japanese system of engineering education with the U.S. system and describe changes in Japanese supply and employment patterns among engineers, so that it becomes clear how these changes will lead to an adjustment in the existing structure of the supply and employment system.

Then, we describe a more detailed survey conducted by the National Institute of Science and Technology Policy (NISTEP). On the basis of this survey, we identify detailed differences between graduates of major universities and the average graduates, as well as differences among engineering subdisciplines.[2]

---

[1] We would like to express our thanks to Mr. M. Kawasaki and Mr. Y. Hirano, who encouraged us to study the issues concerning S&E labor planning.

[2] C. Nishigata, A. Nakanishi, Y. Hirano, "Employment Trends of Science and Engineering Graduates," NISTEP Report No. 1, National Institute of Science and Technology Policy, Science and Technology Agency, June 1989; C. Nishigata and Y. Hirano, "Quantitative Comparison of Science and Engineering Doctorates in Japan and The United States: Training of Researchers in Japanese Doctorate Courses" (in Japanese), NISTEP Report No. 7, National Institute of Science and Technology Policy, December 1989; and



*D. S. Zinberg (ed.), The Changing University*, 101–128.

Based on the data compiled by NISTEP from various sources, we conduct several analyses of how engineering graduates are supplied, how they are employed, and how they are utilized. Specifically, we analyze the following subjects: a unique Japanese system of producing doctoral engineers through dissertations instead of coursework; the growing number of engineers employed by the finance/insurance sector and resulting future economic implications; the rigidity in restructuring universities despite the emergence of new disciplines such as information science; and a case study of engineers in the automotive-manufacturing industry in terms of utilization, assignments, and promotion.

Finally, we try to identify several issues that Japan will have to face in the near future.

## 1. Japanese Supply Patterns of Scientists and Engineers

### *1-1. Major Characteristics of the Japanese System*

The Japanese higher education system of engineers (in institutions above the upper secondary school) consists of four types of institutions: universities, junior colleges, colleges of technology, and special training schools. The university educates the majority of students. Therefore, our analysis of the supply of engineers is concentrated on the graduates of universities.

The main characteristics of the supply pattern of scientists and engineers in Japan are the relatively large number of engineers compared with scientists, and a heavy orientation of students toward bachelor's degrees in both sciences and engineering. Per capita, Japan is producing more engineering bachelors than the United States, while the Japanese production of science bachelors is one-third of the U.S. production. Japan produces less than one-third the doctorates in engineering of the United States, and its production of science doctorates is less than one-sixth of the United States'.

The Japanese university system consists of three types of institutions: national universities funded by the national government, public universities funded by local governments, and private universities. In 1988 national universities graduated 24,000 engineers with bachelor's degrees, public universities graduated 10,000, and private universities produced 76,000. Because national universities and public universities are similar in nature – both types are funded publicly – we will include these two types under the category of national universities. These national universities produced one-third of all engineering bachelors, while private universities produced two-thirds.

The subdiscipline most studied is electrical/electronics engineering, followed by mechanical engineering. These two disciplines comprise 50 percent of all the engineering

---

F. Kodama, "Some Analysis on Recent Changes in Japanese Supply and Employment Patterns of Engineers," in *Science and Technology to Advance National Goals: Science and Technology Policies in the United States and Japan* (Japan Society for Promotion of Sciences, 1989): 41-58.

bachelor's degrees. This concentration partly explains why Japanese industry is enjoying a competitive edge in the technological areas of "mechatronics."[3] On the other hand, the share of applied chemistry is 12.1 percent, and metallurgy's share is only 2.1 percent, which explains why the competitiveness of the Japanese material industry is limited to some areas.[4]

Comparing national and private universities in Japan, we argue that private universities are more concentrated on a few subdisciplines than national universities. Three subdisciplines – mechanical engineering, electrical/electronics, and civil/architecture – comprise a little more than three quarters (76.8 percent) of all the subdisciplines offered by all the private universities, although the concentration is less visible in national universities.

The other differences can be found in the traditional areas of metallurgy, mining, textiles, and shipbuilding, as well as in the new discipline of management engineering. These differences can be explained by two factors. Publicly funded and privately funded universities have different missions: the public universities must make all the subdisciplines available. Also, national universities are older on average than private universities. The Japanese industrial structure had been built around light industries such as textiles before World War II, and coal was a main source of energy. During the recovery period, however, it was restructured toward heavy industries and restructured again toward high technology after the energy crisis.[5] The government established the older national universities when textiles, coal, steel, and shipbuilding were the key industries.

### *1-2. Structural Changes in the Supply of Engineers*

Since 1965, the number of engineering degrees awarded in Japan has increased dramatically. More bachelor's graduates are receiving master's degrees, but the percentage of these graduates who earn a doctorate has not increased. In 1988, Japan produced 76,000 engineering bachelor's degrees and 13,000 science bachelor's degrees, a ratio of 5.7 to 1. The country produced 11,000 engineering master's degrees and 721 doctorates in 1988. Thus, the ratio of bachelor's to master's to doctorates is 100 to 15 to 1. The corresponding ratio for science is 100 to 18 to 4.

Therefore, engineers and scientists are just as likely to proceed further with higher

---

[3] F. Kodama, "Japanese Innovation in Mechatronics Technology: A Study of Technological Fusion," *Science and Public Policy*, 13 (1986): 44-51; and F. Kodama, "Technological Diversification of Japanese Industry," *Science* 233 (1986): 291-296.

[4] F. Kodama, "Can Empirical Quantitative Study Identify Changes in Techno-Economic Paradigm?" *Science, Technology and Industry Review*, No. 7, p 101-129, OECD, 1990.

[5] E. Aminullah, "The Inductive Power of Japanese Technological Innovation: An Empirical Analysis with Special Emphasis on Energy Crisis," M.S. dissertation paper submitted to the Graduate School of Policy Science, Saitama University, Japan, August 1989.

education up to the master's degree level. However, the probability of a science bachelor's

**Figure 1**
**The Growth in Supply of Engineering Graduates**

(Index: 1975 = 100)

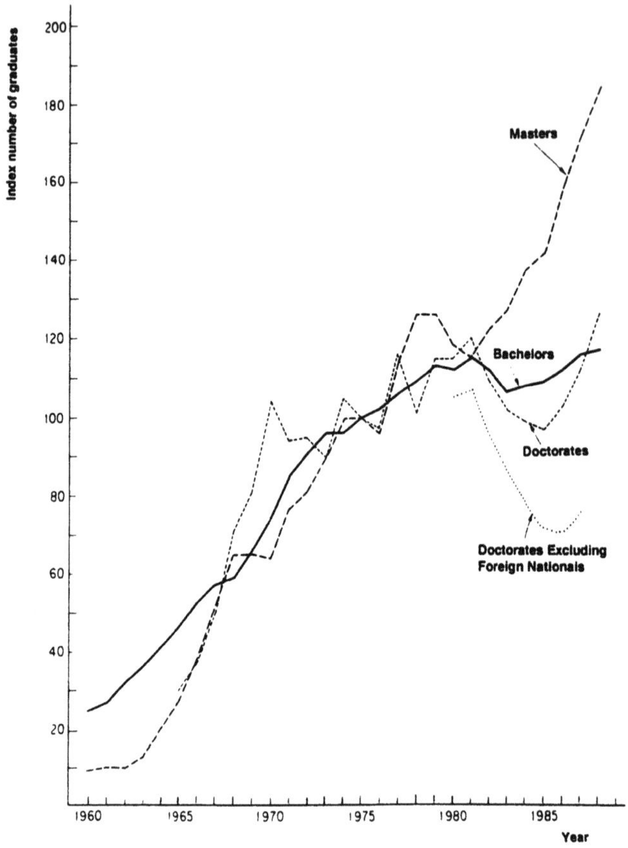


degree recipient entering the doctoral course is four times higher than that of engineering bachelor's degree recipients.The long-term growth in the supply of engineering graduates is depicted in Figure 1. Until 1975 Japan experienced large increases at all levels of education. The supply of bachelor's degrees more than doubled, and master's degrees and doctorates more than tripled. Since 1975, however, the growth rates have slowed in bachelor's degrees and doctorates: we observed a fairly modest increase in bachelor's degrees and a very small increase in doctorates, but master's degrees continue to increase.

The large increase in enrollment during the period 1965-75 is an obvious reflection of the rapid growth of the Japanese economy. The continued increase of master's degrees after 1975 reflects the shift of the Japanese industrial structure from low-technology to high-technology areas. However, we cannot observe any influence of this shift on the increase in doctorates. In fact, as can be seen in Figure 1-1, the number of doctorates granted has continued to decline steadily since 1980, if foreign students are excluded.

### *1-3. Unpopular Engineering Doctorate Program?*

The relatively small number of engineering doctorates and their modest increase are explained by two factors. One is that employers prefer master's degrees over doctorates, because they think master's degree holders are more flexible in adapting to their needs. The other is the strong inclination of students in doctoral courses to take academic jobs. This tendency is demonstrated by Table 1-1, which shows the percentage of doctoral graduates employed by universities and by the manufacturing sector.

*Table 1-1.* Percentages of Doctoral Graduates Employed by University

| | employed by universities | employed by manufacturers |
|---|---|---|
| 1980 | 38% | 40% |
| 1985 | 35 | 34 |
| 1988 | 41 | 28 |

*Source*: Report of Basic Survey on Schools, MESC (Ministry of Education, Science and Culture)

Even in the case of the engineering doctorate, the percentage of those who take university jobs is much higher than that of those who take jobs in the manufacturing sector: in 1988, 41 percent were employed by universities, and only 28 percent by the manufacturing sector. Moreover, the recent saturation in the growth of academic jobs is lowering the employment ratio of the doctoral graduate.

The combined effect of the tendency for the engineering doctoral graduate to take academic jobs and the industry's preference for master's graduates over doctorates results in a lower unemployment figure for master's graduates and a higher figure for doctorates.

More than 20 percent of engineering doctorates were unemployed,[6] while less than 2 percent of engineering master's and bachelor's graduates were unemployed during 1985-88.

*Table 2-1.* Long-Term Trend in Employment Pattern of Engineering Bachelors

| year | total of employed | manufactur-ing | | computer software | | finance/ insurance | |
|---|---|---|---|---|---|---|---|
| | A | B | (B/A) | C | (C/A) | D | (D/A) |
| 1965 | 26,698 | 17,656 | 66% | 189 | 0.7% | 94 | 0.4% |
| 1975 | 54,234 | 27,848 | 51 | 2,712 | 5.0 | 641 | 1.2 |
| 1980 | 62,131 | 31,473 | 51 | 4,121 | 6.6 | 457 | 1.5 |
| 1985 | 59,216 | 35,373 | 60 | 5,881 | 9.9 | 454 | 0.8 |
| 1986 | 60,279 | 35,916 | 60 | 7,086 | 11.8 | 367 | 0.6 |
| 1987 | 61,883 | 36,197 | 58 | 7,713 | 12.5 | 586 | 0.9 |
| 1988 | 61,822 | 32,829 | 53 | 8,611 | 13.9 | 1,193 | 1.9 |

*Source*: same as Table 1-1

## 2. Changing Employment Patterns of Engineering Graduates

### *2-1. Shift of National Economy into Service Industries*

As in every industrialized country, national economic activity is shifting into service industries in Japan. This change affects the employment patterns of scientists and engineers. Table 2-1 shows the long-term trend in the employment of engineering bachelor's degrees. In order to show the influence of this shift on the employment pattern of engineers, certain subcategories such as manufacturing industries, the computer software industry, and the finance/insurance sector are included.

The total employment of engineering bachelor's graduates increased by 14 percent from 1975 to 1988; those employed by manufacturing industries increased only by 18 percent in the same period. However, employment by the computer software industry increased by as much as 185 percent and in the finance/insurance sector by 86 percent.

About half of the engineering bachelor's graduates who joined the work force in 1988 were employed by manufacturing industries, compared with 66 percent in 1965. Japan experienced a drastic decrease in the late 1970s: the percentage reached as low as 45 percent in 1979. This was due to the general economic recession caused by the oil crisis,

---

[6] The employed includes those whose employment conditions include receiving salary, wages, compensation or other forms of employment-related constant income; and those who are self-employed. "Household assistants," however, are considered unemployed.

which every industrialized country experienced. However, Japan experienced a drop in absolute numbers in 1988 – a decrease by more than 3,000 bachelor's graduates from the previous year. It was the first time that the number and the percentage dropped drastically under conditions of good economic growth. Therefore, this drop has more profound causes and implications.

The computer software industry employs as many as 15 percent of all engineering bachelor's graduates who have jobs, although it comprised only one percent in 1965 and six percent in 1975. A steady increase in this percentage can be observed since 1965. We can interpret this increase as a reflection of the shift of the national economy toward the service sector.

The most recent phenomenon is the drastic increase in employment by the finance/insurance sector: employment more than doubled in 1988 from the previous year. Although the level still remained as low as two percent and there was an increase at that level before 1980, job assignments for engineering bachelor's graduates changed in an essential way: in the past they were employed as specialists involved with computer installation. Now they are employed as generalists with chances to be promoted to corporate managers.[7]

*2-2. Differences among Universities: Major vs. Average*

In order to see if there is any difference in employment patterns between graduates of major universities and the average (depicted in Table 2-1) and any differences among various scientific disciplines, the National Institute of Science and Technology Policy (NISTEP) conducted a detailed survey of 52 departments in ten major universities: six national universities and four private ones; 637 science graduates and 3,747 engineering graduates.[8]

NISTEP found that the tendency to be employed by manufacturing industries is stronger in graduates of the major universities than on average: 66 percent of bachelor's degree holders from the major universities are employed by manufacturing industries, while only 53 percent of them are employed by manufacturing industries on average. However, the changes in the past three years are more dramatic in major universities than on average: the percentage decreased by ten percentage points from 76 to 66 in major universities, while it decreased by seven percentage points from 60 to 53 on average. The computer software industry employs about the same number of engineering bachelor's graduates from major universities (12 percent) as on average (14.5 percent). Therefore, the shift of employment into the computer software industry is a general and overall trend and will be a long-lasting phenomenon. However, students in major universities are

---

[7] Y. Baba and S. Takai, "Opening the Black Box of Japanese Services Techniques: Information Technology Introduction in the Big Banks," Minemo in National Institute of Science and Technology Policy, December 1989.

[8] NISTEP Report No. 1, op. cit.

slightly less inclined than on average to take a jobs in the computer software industry.

A bigger difference is in the finance/insurance sector. The percent of bachelor's graduates taking jobs in this sector is 4.8 percent from major universities and only 1.9 percent on average. The increase in the past three years is more dramatic in major universities than on average – it more than tripled in major universities from 1.5 percent to 4.8 percent, while the average ratio doubled from 0.8 percent to 1.9 percent.

Therefore, we can speculate that larger finance/insurance institutions who employ science and engineering bachelor's graduates are targeting their recruitment toward major universities, and students in major universities are more interested in being employed by the finance/insurance sector.

### 2-3. *Differences among Engineering Subdisciplines*

NISTEP's survey divides engineering disciplines into three categories: mechanical, electrical/electronics, and materials (which includes only metallurgy and applied chemistry). Manufacturing industries are still the majority employee for these three disciplines, as shown in Table 2-2.

*Table 2-2.* Disciplinary Differences in Employment Ratios by Manufacturing Industries ( ): the ratios of master's graduates

| year | mechanical engineering | electrical/ electronics | materials engineering |
|---|---|---|---|
| 1986 | 80% (89%) | 71% (73%) | 76% (95%) |
| 1987 | 77 (76) | 66 (67) | 74 (90) |
| 1988 | 71 (75) | 61 (61) | 65 (69) |

*Source*: C. Nishigata, A. Nakanishi, Y. Hirano, "Employment Trends of Science and Engineering Graduates," NISTEP Report No. 1, National Institute of Science and Technology Policy, Science and Technology Agency, June 1989.

As seen in the table, the percentage of bachelor's graduates employed by manufacturing industries is highest in mechanical engineering at 71 percent and lowest in electrical/electronics engineering at 61 percent. However, overall the ratios show a constant decline over time throughout all the disciplines. Moreover, it is noteworthy that the difference in ratios between bachelor's and master's degrees has lessened: the differences are reduced to less than five percent in 1988, down from 20 percent in 1986.

Among engineers with bachelor's degrees employed by the computer software industry, there is little difference in the disciplines studied (Table 2-3).

All the figures in the table are more or less in the range of 10 percent: computer employment is highest in materials engineering and lowest in mechanical engineering. However, the computer software industry is gradually and steadily employing more bachelor's graduates of all the disciplines.

In employment by the finance/insurance sector (Table 2-4), marked disciplinary differences can be found.

*Table 2-3.* Disciplinary Differences in Employment Ratios by the Computer Software Industry
( ): the ratios of master's graduates

| year | mechanical engineering | electrical/ electronics | materials engineering |
|---|---|---|---|
| 1986 | 8 (4) | 9 (7) | 9 (2) |
| 1987 | 10 (6) | 13 (10) | 11 (2) |
| 1988 | 11 (6) | 14 (9) | 10 (7) |

*Source*: same as Table 2-2

*Table 2-4.* Disciplinary Differences in Employment Ratios by the Finance/ Insurance Sector, ( ): the ratios of master's graduates

| year | mechanical engineering | electrical/ electronics | materials engineering |
|---|---|---|---|
| 1986 | 1.5 (0.0) | 0.9 (0.7) | 1.1 (0.0) |
| 1987 | 1.7 (0.3) | 1.5 (0.6) | 2.9 (1.0) |
| 1988 | 4.3 (1.5) | 2.6 (1.6) | 7.0 (2.5) |

*Source*: same as Table 2-2

For materials engineering, bachelor's graduates employed reached seven percent, 2.6 percent in electrical/electronics engineering and 4.3 percent in mechanical engineering. Therefore, employment by the finance/insurance sector is not associated with the specific nature of the disciplines. A more plausible hypothesis might be that the less attractive a discipline is to able engineering students, the more likely they are to choose employment in the finance/insurance sector. This phenomenon is good evidence that engineering bachelor's graduates are being employed as generalists.

## 3. Dissertation Doctorates: Another Japanese Miracle?

It is well known worldwide that the Japanese economy grew very rapidly, industrial research and development activities grew at a higher rate than the economy as a whole, and Japanese exports shifted toward higher value-added products. Also, it became clear that Japanese gains in competitiveness were in high-technology products, as reported in

Narin and Olivastro.[9] These phenomena have been called an "economic miracle" by Westerners. However, the increase in engineering doctorates was very marginal. Therefore, it seems that the Japanese economic miracle was accomplished without raising the doctoral educational level – a puzzle in higher education.

First, let us formulate this puzzle. The ratios of students proceeding to advanced programs of engineering education are shown in Table 3-1.

*Table 3-1.* Proceeding Probabilities to Higher Engineering Programs

| | master's | doctorates |
|---|---|---|
| year | bachelor's | master's |
| 1965 | 9.5% | 32.8% |
| 1975 | 10.1 | 11.3 |
| 1980 | 9.8 | 7.8 |
| 1985 | 13.9 | 8.3 |
| 1986 | 14.3 | 9.3 |
| 1987 | 15.1 | 8.4 |
| 1988 | 16.1 | 8.9 |

*Source*: same as Table 1-1

The probability that bachelor's graduates proceed to master's degrees increased rapidly: the probability is currently 16 percent, compared with ten percent in 1965. On the other hand, the probability that master's graduates proceed to doctorates decreased drastically in the past decade: it is now only nine percent, while it was 33 percent in 1965. These two probabilities behaved in opposite directions.

The situation in doctoral programs is much worse than it appears. According to Japanese statistics, graduation from doctoral programs does not necessarily mean that students received doctoral degrees. In those statistics, "doctoral graduates" refers to those who:

(1) have completed the required period of academic residence;
(2) have met the required academic units; and
(3) have been judged as having met the required level of academic standards as determined by a dissertation or final examination evaluation.

---

[9] F. Narin and J. Frame, "The Growth of Japanese Science and Technology," *Science* 245 (1989): 600-605; and G. Bylinsky, "The High-Tech Race: Who's Ahead?" *Fortune* (October 13, 1986): 18-36.

However, the statistics also state that one may be considered a doctoral graduate if only the first two conditions have been met. Therefore, many graduates from doctoral programs leave these programs without receiving a doctoral degree. About 30 percent of doctoral students leave their programs without a degree, although the situation has improved in recent years. In 1980 as many as 44 percent of doctoral students left their programs without a degree. Although some improvements can be observed in the national universities, which retain a dominant position in the supply of doctorates, graduation without a degree is the norm in private universities: on average in the period 1975-88, more than 50 percent of the graduates are without doctoral degrees. There is no improvement visible, and in fact the situation has begun to deteriorate again over the past two years.

Moreover, foreign students are the majority in doctoral programs in several universities. In 1987 the percentage of engineering doctoral students who are foreign had already surpassed 30 percent. It is still far below the percent of U.S. engineering doctoral students, which is close to 50 percent, but is well over that of U.S. science doctoral programs.

It is undeniable that fewer foreign doctoral graduates are employed in Japan compared with the United States. Therefore, the phenomenon of fewer native doctoral students is more damaging in Japan, and poses a more serious problem to the Japanese taxpayer, since most engineering doctoral programs are offered by national universities in Japan.

The answer to this puzzle might be found in the system called "dissertation doctorates." In Japan there are two ways to earn a doctorate. One is the ordinary path, in which a student does coursework and fulfills the requirements imposed by a university. By the other route, an applicant submits an acceptable dissertation paper to the university. Many researchers in industry earn degrees through the latter method, that is, researchers working for companies write papers based on their company work. In Table 3-2, the number of doctorates granted by dissertation is shown with those by coursework. In 1986 approximately 1,000 doctorates were granted through dissertations, while 500 doctorates were earned through coursework. In other words, the dissertation doctorate appears to be the major route to earn a doctorate in Japan.[10]

As early as 1963 the number of doctorates granted through dissertations surpassed those granted by coursework. Until 1982 the number of dissertation doctorates stayed more or less equivalent to the number of coursework doctorates. However, the number of dissertation doctorates began to increase rapidly after 1983, while that of coursework doctorates decreased slightly. Therefore, the dissertation doctorate has become the dominant mode of granting engineering doctorates, representing almost two-thirds of them.

[10] F. Kodama, "Direct and Indirect Channels for Transforming Scientific Knowledge into Technological Innovations," in *Transforming Scientific Ideas into Innovations: Science Policy in the United States and Japan*, eds. B. Bartocha and S. Okamura (Tokyo: Japan Promotion for the Promotion of Science, 1985) p. 198-204; and F. Kodama, "Dissertation Doctors Increase Rapidly As Japan Gains Edge in Technology," *The Japan Economic Journal* 2 (September 1989).

It is difficult to compare the quality of dissertation doctorates versus coursework doctorates. All we can do is measure the distribution of dissertation doctorates between national and private universities and between major and average universities.

*Table 3-2.* Number of Engineering Doctorates Granted

| year | coursework | dissertation | share of dissertation |
|---|---|---|---|
| | A | B | B/(A+B) |
| 1960 | 72 | 6 | 8% |
| 1961 | 69 | 17 | 20 |
| 1962 | 84 | 54 | 39 |
| 1963 | 94 | 116 | 55 |
| 1970 | 425 | 428 | 50 |
| 1980 | 523 | 663 | 56 |
| 1981 | 541 | 695 | 56 |
| 1982 | 506 | 772 | 60 |
| 1983 | 489 | 801 | 62 |
| 1984 | 447 | 844 | 65 |
| 1985 | 480 | 924 | 66 |
| 1986 | 505 | 988 | 66 |
| 1987 | 621 | 926 | 60 |

*Source*: C. Nishigata and Y. Hirano, "Quantitative Comparison of Science and Engineering Doctorates in Japan and the United States: Training of Researchers in Japanese Doctorate Courses," (in Japanese) NISTEP Report No. 7, National Institute of Science and Technology Policy, December 1989.

The national university plays a dominant role among engineering doctorate programs. National universities grant roughly the same percentage of dissertation doctorates as coursework doctorates; they grant 86 percent of the dissertation doctorates and 85 percent of coursework doctorates.

Major national universities grant more dissertation than coursework doctorates, while private universities grant more coursework doctorates. Major national universities granted as many as 97 percent of all the dissertation doctorates granted by national universities, and 91 percent of coursework doctorates granted by national universities. In the case of private universities, 46 percent are granted by major private universities, while the corresponding figure for coursework doctorates is 56 percent. If we suppose that major national universities are more demanding in their criteria for doctoral degrees than other universities, then we can assume that the quality of dissertation doctorates is not lower than that of coursework doctorates and is probably a little higher.

Therefore, it is necessary to include dissertation doctorates in statistics about the

Japanese production of doctorates. Including dissertation doctorates, the increase in the rate of doctorates is approximately equivalent to that of master's degrees, as shown in Figure 2. It is obviously far more than that of bachelor's degrees. In fact, the number of Japanese engineering doctorates exceeds that of the United States on a per-capita basis, if foreign nationals with temporary visas are excluded. Therefore, we can conclude that there is no miracle: the Japanese economic miracle was accomplished with a corresponding enhancement of Japan's educational level.

*Table 3-3.* Per-Capita Research Expenditure, by Types of Organization (million yen)

| year | university | company | difference | |
|---|---|---|---|---|
| | (A) | (B) | (B/A) | (B-A) |
| 1970 | 3.94 | 8.75 | 2.22 | 4.81 |
| 1975 | 6.30 | 11.49 | 1.82 | 5.91 |
| 1980 | 8.18 | 18.14 | 2.22 | 9.96 |
| 1985 | 9.11 | 25.70 | 2.82 | 16.59 |
| 1986 | 9.25 | 24.31 | 2.63 | 15.06 |
| 1987 | 9.74 | 24.90 | 2.56 | 15.16 |

*Source*: White Paper on Science and Technology 1989, Science and Technology Agency.

One might interpret the drastic increase in dissertation doctorates over coursework doctorates as a reflection of the university's inability to attract young researchers with its teaching capability and research facilities' quality. As an indicator of research environment, the per-capita research expenditure is compared between the university and the company as shown in Table 3-3.

The gap between the university and the company is widening. One interpretation is that the age of high technology demands the dissertation method of producing engineering doctorates, because the speed of innovation is so fast that knowledge accumulated in the university becomes obsolete too soon.[11] A more crucial reason might be that this is the only available and efficient method by which young researchers in engineering have an

---

[11] F. Kodama, "Research and Development Dynamics of High-Tech Industry: Toward the Definition of High Technology," *Journal of Science Policy and Research Management* 1 (1986): 65-74; and F. Kodama, "High tech developments render traditional management obsolete", *The Japan Economic Journal* 26 (October 1988).

opportunity to tackle real problems. If the latter is the case, the dissertation path to the doctorate will become the standard method of producing engineering doctorates in Japan.

**Figure 2**
**Engineering Degrees Including Dissertation Doctorates**
(Index, 1975 = 100)

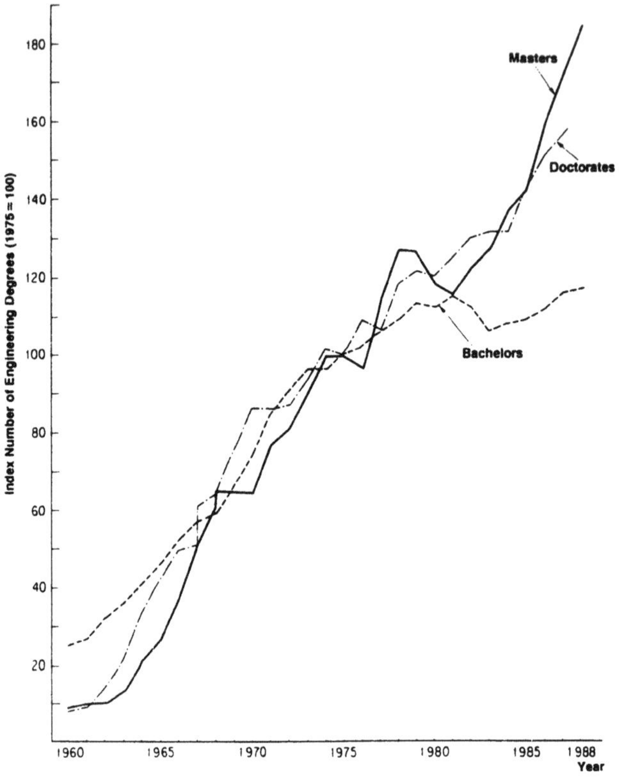


However, even if it is a good educational method, research is a different problem. What will become of university research without doctoral students? The university will not be able to function as a center of research excellence.[12] The problem becomes more serious since the average age of dissertation doctorate recipients is much older than that

---

[12] F. Kodama, "Access to Japanese Basic Research: Its Centers of Activity and of Excellence," in *Proceedings of International Symposium on International Cooperation and Competition in Science and Technology* (Engineering Academy of Japan, April 1988): 52-58.

of coursework doctorate recipients and is getting older.

In 1988 the average age of dissertation doctorate recipients was 42, while that for coursework doctorates was 29. From 1970 to 1988, the age distribution skewed toward the older percentile. Counting just the dissertation doctorates who are younger than 35 years old, we see that in 1960, three-quarters of doctorates were granted to young people, while less than one quarter were granted to young people in 1988.

## 4. Employment by the Finance Sector: Are We Losing or Gaining?

### *4-1. A Paradox of Income Distribution*

One of the obvious reasons for the major changes in the employment patterns of scientists and engineers derives from changes in the demand structure in the past few decades. As can be seen in Table 4-1, employment by the tertiary sector doubled since 1965, while that in the secondary sector increased only 1.3 times.[13]

*Table 4-1.* Number of Employees by Sectors
unit: ten thousands; ( ): the composition ratio

| year | primary industry | secondary industry | tertiary industry |
|---|---|---|---|
| 1965 | 63 (2) | 1,266 (46) | 1,455 (52) |
| 1975 | 46 (1) | 1,530 (42) | 2,069 (57) |
| 1980 | 45 (1) | 1,572 (40) | 2,352 (59) |
| 1985 | 43 (1) | 1,655 (38) | 2,607 (61) |
| 1986 | 44 (1) | 1,652 (38) | 2,675 (61) |
| 1987 | 44 (1) | 1,635 (37) | 2,744 (62) |
| 1988 | 45 (1) | 1,688 (37) | 2,796 (62) |

*Source*: The Labor Force Survey, Statistics Bureau, Management and Coordination Agency.

The secondary sector share of employees decreased from 46 to 37 percent, while the tertiary sector increased from 52 to 62 percent. Since demand in the job market reflects

[13] The primary sector includes agriculture, forestry and fisheries. The secondary sector includes mining, construction and manufacturing. The tertiary sector includes utilities, transport and communication, wholesale and retail trade, financing and insurance, services, and government.

net increases in employment opportunity, it is reasonable to assume that university graduates are being absorbed by the tertiary sector.

Thus, we can also assume the absolute and relative increase in the employment of scientists and engineers by the finance/insurance sector is related to the widening gap in income between employees of manufacturing industries and that of the finance/insurance sector.

In Japan, pay is more or less determined by the employee's age. At 35, university graduates in the finance/insurance sector are paid 30 percent higher salaries than those in manufacturing industries. In annual income, which includes both salary and bonus, the gap is widened to 45 percent. In 1976 these differences were 15 percent and 27 percent, respectively. The gap almost doubled in the past 10 years – and is still widening. In 1976 the gap peaked at age 52 in both absolute and relative terms: the difference was as much as 2.7 million yen, and it amounted to 37 percent of annual income in manufacturing industries. In 1986, however, it peaked at age 43 in absolute terms: the difference was 2.7 million yen. It peaked at the age of 35 in relative terms: the difference amounted to 46 percent of annual income in manufacturing industries. In other words, the gap is widening and becoming effective earlier in the careers of engineers.

If this phenomenon of a widening income gap is related to the growth in the manufacturing production index, we seem to be developing a paradox of income distribution: the more the manufacturing industry becomes prosperous, the poorer its employees become compared with those in the finance/insurance sector, whose prosperity is built on that of manufacturing industries. We can think of several explanations. Since competition among the manufacturing companies is getting tougher and tougher, they have to use their profits to modernize their manufacturing facilities and invest in research and development to survive the high-technology race. On the other hand, the finance sector is not facing such competition, because it is still fairly protected and regulated in Japan.

A more profound and intrinsic reason can be found in the unique characteristics of the Japanese wage structure. It is not easy for the manufacturing industry to raise the wage rate of the engineering university graduate because of the lifelong employment system and the egalitarian wage system, which has no debilitating status or salary differentials between white-collar and blue-collar workers or between university-graduate engineers and production-line workers.

The percentage of university graduates in the finance/insurance sector is as high as 90 percent of all employees, while it remains as low as 30 percent in the manufacturing industry. The percentage of university graduates increased from 40 percent to 90 percent in the past 20 years in the finance/insurance sector. On the other hand, in the manufacturing sector it reached saturation at a level of 30 percent in the past 15 years. In this situation, an increase in wages for university graduates would inevitably cause substantial cost increases in the manufacturing industry, while this is not the case in the finance/insurance sector.

*4-2. Are We Losing or Gaining Precious Talents?*

A question arises from the drastic and seemingly long-lasting increases in the employment of engineering graduates by the finance/insurance sector: Are we losing precious talent to sectors outside manufacturing, or is the country as a whole gaining talent as service sectors take the leading position?

Obviously, manufacturing is losing much of the talent it definitely needs. However, it is not the most rapidly growing sector nor is it the most profitable. The finance/insurance sector shows the most growth and will become more internationalized. Moreover, its nature has entirely changed, mainly because of information and communication technologies, the technological innovations produced by the manufacturing sectors.

A recent study for the U.S. Congress by the Office of Technology Assessment described the scope of this influence:

> The financial service industry is markedly different from what it was at the end of the 1970s, and the rate of change will only slow slightly during the remainder of the 1980s. Advancing information and communication technologies are key factors that have changed the nature of financial services: the applications of advanced information and telecommunication technologies in systems for delivering financial services change the way those services are created, delivered, priced, received, accepted, and used. Relationships between and among users and providers of financial services are changing.
>
> Communication will be the key to delivering financial services in the future. Networks growing out of those used to connect shared systems of automated teller machines are likely to provide the basis of systems permitting electronic initiation of fund transfers from the merchant's counter. Systems providing access to funds from virtually any place in the nation regardless of where they are deposited are now being developed and are likely to be in use in the next few years. Advanced communication technologies including satellite relays, video cable, fiber optics, and cellular radio will find wide application in the financial industry.
>
> Decreasing computer costs will create the opportunity for large number of individual consumers and managers of small businesses to take advantage of technology in using financial services. Large computers will be used to support the data bases and the communication processing needed to operate the large, interactive financial service delivery systems of the future. Computers that accept voice inputs and recognize fingerprints may become cost effective for financial service delivery systems by the turn of the century. Small, inexpensive personal computers in both home and office will make it possible for customers to interact with a multiplicity of financial service

offerors. Computer processor and memory chips embedded in plastic cards may find widespread use in the financial service industry.[14]

In other words, jobs in the financial sector have drastically changed: the financial service industry is now thinking about optimal ways of operating the system and developing new industrial products in financial markets. John Reed, chairman of Citicorp, stated that the role of major financial institutions in stimulating technological innovations is not limited to the simple lending of funds. In the broadest sense, it lies in creating new financial techniques or in innovating financial technologies.[15] It might not be so different from what the manufacturing sectors search for in efficient factory operation and the innovation of new products.

What then should be the essence of an engineering education? One possible answer is that an engineering education should teach students how to integrate all available knowledge for specific purposes. This type of training can be transferred across sectoral borders, i.e., from the manufacturing to the service sector. If that is the case, the best method of teaching integration might be in courses about hardware, because more teaching materials have been accumulated, it is visible, and easier to understand.

### *4-3. Response of Banks and Students*

The changes occurring in bank businesses are inducing a drastic change in both demand and supply of engineers. We found that science and engineering graduates are more concentrated in the large financial institutions than are graduates of other disciplines.

As many as 80 percent of the science and engineering graduates in the finance sector are employed by the 22 largest financial institutions, while the corresponding figure for graduates in the other disciplines is as low as 45 percent. Therefore, we can conclude that the institutions that are leading in financial technology innovation are recruiting science and engineering graduates.

An important supply-side question is whether engineering students from prestigious universities are more motivated to be employed by banking institutions.

Because of the severe competition in entrance examinations for several prestigious universities, the tutoring industry, which gives high-school students lessons in addition to their formal classes, has grown in Japan. Most students who will go to universities attend these classes. The tutoring industry has developed a fairly reliable system in which almost all the departments in almost all Japanese universities can be ranked in the difficulty of their entrance examinations, roughly equivalent to the SAT (Scholastic Aptitude Test) in

---

[14] Office of Technology Assessment, "Effects of Information Technology on Financial Services Systems," OTA-CIT-202 (Washington, D.C.: U.S. Government Printing Office, 1984).

[15] J. Reed and G. Moreno, "The Role of Large Banks in Financing Innovation," in *The Positive Sum Strategy: Harnessing Technology for Economic Growth*, eds. R. Landau and N. Rosenberg (National Academy Press, 1986), pp. 453-465.

the United States.

Based on this database of the Japanese SAT, we conducted a preliminary analysis of the relation between a university's SAT scores and its students' tendency to take jobs in banks. Nine engineering schools were chosen as a sample, of which six are national universities and three are private universities. However, no bachelor's graduates from one of the private schools were employed by banks in 1988. Therefore, the remaining eight universities were selected for our statistical test. The result is shown in Figure 3.

**Figure 3**

**Relation between SAT and Employment by Banks**

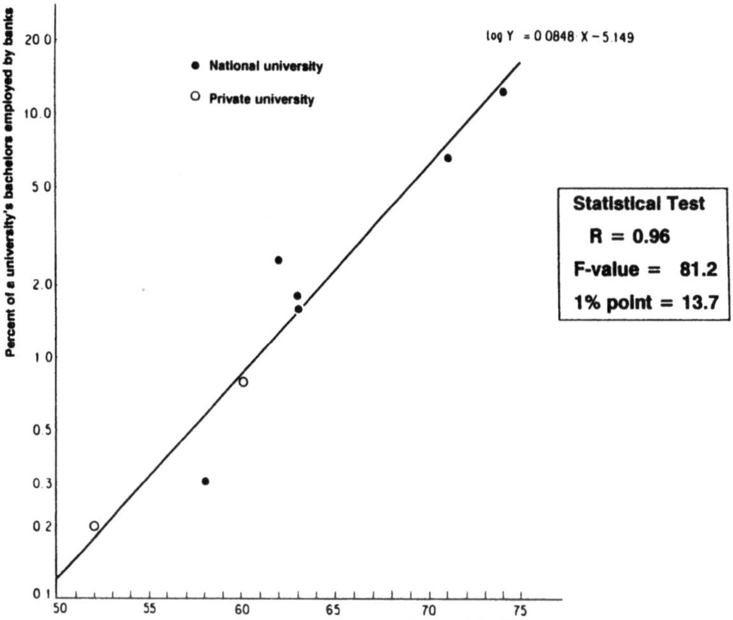


As seen in the figure, we tried a statistical fit in the following log-linear relation:

$$\log Y = a \cdot X + b$$

where,

X: a university's SAT score,

Y: the probability that a student will be employed by a bank.

We obtained a correlation coefficient as high as 0.96. The F-value is as high as 81.12, far above 13.7, i.e., one percentage point of F-distribution. This analysis clearly shows that the higher a university's SAT score, the higher the probability that a university's engineering bachelor's graduates will take jobs in banks.

### *4-4. Are We Developing a Myopic System?*

A historical analysis of leading Japanese industries tells us about the interaction between the recruitment of engineering bachelor's graduates and the internationalization of these industries:[16] whenever an industry is going to grow and internationalize, young engineering talents are mobilized and push the internationalization forward. This pattern can be found in the textile industry, the shipbuilding industry, trading companies, and the automobile industry. Why will this not happen in the finance/insurance sector?

An early indication can be found in an international comparison of annual productivity growth measurements of selected sectors made by Gordon and Baily, as shown in Table 4-2.[17]

*Table 4-2.* Average Annual Growth in GDP per Worker in the Manufacturing and Finance/Insurance Sectors (Percent per year)

| | Manufacturing | | | Finance/Insurance | | |
|---|---|---|---|---|---|---|
| | Japan | U.S. | gap | Japan | U.S. | gap |
| | A | B | (A-B) | C | D | (C-D) |
| 1973-1979 | 5.21 | 1.03 | (4.18) | 1.96 | -1.01 | (2.97) |
| 1979-1985 | 6.02 | 3.58 | (2.44) | 1.77 | -2.01 | (3.78) |

*Source*: R. Gordon and M. Baily, "Measurement Issues and the Productivity Slowdown in Five Major Industrial Countries," presented to the International Seminar on Science, Technology and Economic Growth, OECD, Paris, June 5-8, 1989.

In the finance/insurance sector, Japan has recorded positive figures in contrast with the U.S.'s negative ones during 1973-85. More important, the difference between Japan and the United States has widened from 2.97 percent between 1973 and 1979 to 3.78 percent between 1979 and 1985. On the other hand, the difference has narrowed from 4.18 to 2.44 in the manufacturing sector.

However, we have to ask a basic question: Is prosperity in the finance/insurance sector

---

[16] F. Kodama, "Dynamic Interactions between Technology Transfer and Engineering Education," in *Proceeding of the Fourth U Science Policy Seminar on Engineering Education: United States and Japan*, ed. E. David and T. Mukaibo (Washington, D.C.: National Science Foundation) p. 187-193.

[17] R. Gordon and N. Baily, "Measurement Issues and the Productivity Slowdown in Five Major Industrial Countries," presented to the International Seminar on Science, Technology and Economic Growth, OECD, Paris, June 5-8, 1989.

possible without an equivalent prosperity in manufacturing? Even if it is possible, is it healthy from the viewpoint of the national economy? Its effect might be more than just removing precious talent from the limited pool of the able young population.

On the basis of British experiences, K. Pavitt and P. Patel come to the following conclusion concerning the interaction between financial markets and technical innovation:

> The essential distinction that we make is between national systems that are technologically dynamic, and those that are technologically myopic. For our purposes, the former recognize the cumulative, irreversible, and uncertain nature of technological activities, whereas the latter do not.
>
> In myopic systems, the technological activities are evaluated like an ordinary investment, namely, on the basis of their prospective rate of return in responding to an existing and precise market demand. In dynamic systems, the evaluation of technological activities also includes the prospect of creating new market demands, and of accumulating over time knowledge and experience that open up further technological applications and business opportunities in future.[18]

The efficient and perfect operation in financial markets might discourage technology development, because investments are biased toward short-term profits, and risk taking in uncertain environments such as technology developments is discouraged. At least so far, the Japanese investment patterns in high-technology industries have not been guided by the rate of return.[19]

## 5. Emergence of New Disciplines: Who Is Adjusting?

As we described above, both the economic structure and the employment patterns of science and engineering graduates have changed drastically. This economic structural change created several new industries, such as the computer software companies. Within this new industry, the demand for graduates of higher education has also increased. Degree engineers had composed a minority in 1965; however, they were a little more than two-thirds of all engineers employed by the computer software industry during the 1970s. This number has reached more than 80 percent in 1988. The change in the composition of engineers is a clear manifestation of the transformation of the computer software industry from labor-intensive to knowledge-intensive and from lower value-added to higher value-added. This also indicates that the industry demanded new educational programs, such as software engineering or probably a new engineering subdiscipline.

---

[18] K. Pavitt and P. Patel, "The International Distribution and Determinants of Technological Activities," *Oxford Review of Economic Policy* 4 (1988): 51.

[19] F. Kodama, "How Research Investment Decisions Are Made in Japanese Industry," in *The Evaluation of Scientific Research*, edited by Ciba Foundation (John Wiley & Sons, 1989) p. 210-214.

Therefore, the last question to be addressed is whether university education has responded to this change.

Since the economic structural changes and the changes in employment patterns we have described are related to the emergence of information technology, we chose the establishment of information-science departments as an indicator of the responsiveness of university education to Japan's changing needs. The educational capacity of information science can be measured by the number of students who can be enrolled in departments of information science.

The enrollment capacity was 6,820 students in 1988, comprising only 6.4 percent of the total capacity of education in science and engineering. National universities provided a capacity of 3,830 students, comprising about 10 percent of the total capacity of national universities in science and engineering. Private universities provided a capacity of 2,990, or 5.3 percent of the total capacity of private universities.

However, the percentage had remained below five percent until 1985. National universities responded a bit earlier than private universities, which surpassed five percent in 1988. If the computer software industry were to try to employ only graduates of information-science departments, this percentage would be far below 10.7 percent, the percent of engineering bachelor's graduates employed by the computer software industry in 1985, as shown in Table 2-1.

We can argue that the statistics presented above indicate that universities are not responding to the changing demands; this unresponsiveness is due to their rigid departmentalization. Decisions for the Ministry of Education to create new departments are made more or less based on the priority given to this activity in the budget requests set by the faculty of each individual department in each university. Because of the immobility of these faculties and the inbreeding of teaching staffs within departments, there is less incentive for faculty to decide on the creation of new departments far apart from their own disciplines. Therefore, it is clear that students rather than professors are adjusting to the changing needs of the economy.

There is another way to solve the problem. Although the number of information-science departments is far below those required by the job market, this subject is being taught within existing hardware-related departments.

As we described before, we can think that the essence of an engineering education is teaching students how to integrate all available knowledge for specific purposes. We argued that the best method of teaching integration might be to teach it in the context of the hardware, because more teaching materials have been accumulated than in the other disciplines, and real experience exists. Furthermore, these teaching materials are visible to students and thus make the subject easier to understand. Also, this type of training can be transferred across sectoral borders, i.e., from the secondary industry to the tertiary industry.

In fact, the computer software industry has not had any difficulty in recruiting engineering graduates. It has drawn the needed talents from various science and engineering subdisciplines. The gap between demand and supply has been filled by graduates who learned the subject within hardware-related departments.

This might have been a positive arrangement in a time when the application of computer technology was limited to industrial products, such as machine tools. In order to produce effective computer control programs, the programmer had to know how the machine tool is operated and manufactured. Which pattern would be better – mechanical engineers who learn computer programming by themselves to produce control programs for machine tools, or computer scientists performing these tasks with the help of machine tool designers?

Let us formulate the questions: When a new discipline such as information science appears, should we teach this subject in the existing framework, or should we establish a new and independent department? Which is more effective and efficient? What are the limitations of the former method when the new discipline progresses further?

Some past Japanese experience seems to bear on these limitations. The computer programs developed by hardware engineers who acquired programming skills on their own might work efficiently when use is limited to internal purposes or when the program is built into the hardware. However, hardware producers find it difficult to turn their programs into commercial products to be used for more general purposes.

As for teaching a new discipline such as information science within the framework of existing departments, faculty members who are teaching information science in these departments are, in many cases, people who have never received a formal education in information science. This method might have to end when information science progresses further and unique expertise becomes indispensable.

An early warning can be found in our preliminary analysis, which tries to correlate a university's SAT score to its students' inclination to take jobs within the computer software industry. We used the same samples that were used for the analysis of banks. The results are shown in Figure 4, in which we conducted a statistical fit in the following linear relation:

$$Y = a \cdot X + b$$

where,

X: a university's SAT score,

Y: the probability that a student will take a job in a computer software company.

We obtained a correlation coefficient as high as 0.87. The F-value is as high as 21.69, far above 12.2, i.e., one percentage point of F-distribution. Therefore, this analysis suggests the higher a university's SAT score, the lower the probability of a university's engineering bachelor's graduates will take jobs in the computer software industry. This reveals that the computer software industry is definitely not the first choice for able engineering bachelor's graduates. One of the reasons is the university's rigidity in the way it creates new departments. If the subject of information science were taught in a more systematic way and by more experienced professionals, an arrangement that would provide able engineering students with a chance to study and understand this new discipline better, they might be more motivated to work for the computer software industry, which will definitely become a key industry in the near future.

If the university stays rigid in reorganizing the engineering education system, the situation will further deteriorate for manufacturing industries, for which an excellent computer software industry will become more indispensable.

Figure 4

**Relation between SAT and Employment by Computer Software**

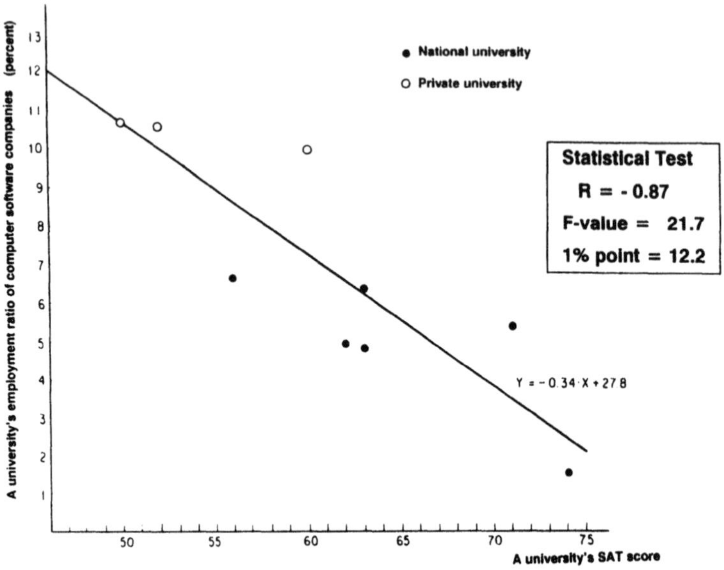


Moreover, the future growth in banking businesses will erode the situation further. We found that a trade-off relation holds between the employment in the banking sector and that in the computer software industry. In order to test this hypothesis, we combined the two SAT analyses described above. The result is shown in Figure 5, in which we conducted a statistical fit in the following linear relation:

$$Y = a \cdot X + b$$

where,

Y: the probability that a university's students will take jobs in computer software companies.

X: the probability that a university's students will take jobs in banking institutions.

We obtained a correlation coefficient of 0.78. The F-value yields 11.2, a little below 12.2: one percentage point of F-distribution, but far above 5.99, its five percent point. Therefore, this analysis suggests a possibility: the higher the probability of a university's engineering bachelor's graduates to be employed by banks, the lower the probability of their being employed by the computer software industry. This relation indicates that the number of engineering bachelor's graduates who are willing to take jobs outside manufacturing industries is limited and is not increasing as rapidly as expected by these two service sectors. In other words, the computer software industry is competing with the finance/insurance sector in recruiting engineering graduates from a limited pool. These two sectors might be in danger of playing a zero-sum game in the near future.

Figure 5

Trade-off between Banks and Computer Software Industry

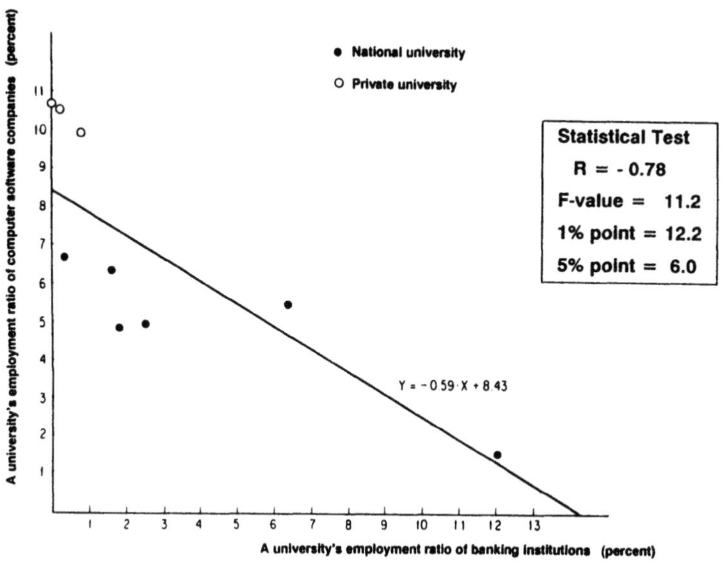


The Federation of Economic Organizations (Kei-Dan-Ren) estimates that Japan will have a shortage of as many as 168,000 software engineers in 1993 and that its existing institutions will meet only 46.2 percent of the future demand of 364,000 software engineers to be needed in 1993.[20] This situation is often termed a "software crisis" worldwide.

Given the scope of the situation, the future supply of software engineers might be a formidable problem whose solution cannot be left only in the hands of engineering educators, and it may be beyond the capacity of engineering schools within the universities. A possible solution might be to expand the supply base of information scientists. This way of thinking is reflected in the recommendation by the Kei-Dan-Ren, which mentioned the reform of universities for the first time in its history and proposed

---

[20] Federation of Economic Organizations, "The Report on the Problems of Employment/Education in the New Economic/Industrial Structure," Keizai-Dantai-Rengoukai, June 27, 1989.

creating a school of information policy in the university.[21] It recommended this new school be established beyond the disciplinary boundary between natural science/engineering and social/humanity sciences.

## 6. Utilization of Engineers: Are They Respected or Exploited?

So far we have described the changing pattern of supply and employment of university-graduate engineers. We have not mentioned how they are utilized after they are employed by the company, because we have not yet conducted a systematic investigation concerning job assignments given to them. However, there is an unpublished survey that investigates the allocation of engineers in the automobile industry, as shown in Table 6-1.[22]

*Table 6-1.* Degree Engineers and Technical Support Personnel Employed in Automotive Manufacturers

| | Japan | U.S.A. |
|---|---|---|
| Percent of Auto Employment: | | |
| Engineers | 7.6% | 4.1% |
| Technical Support Personnel | 6.2% | 2.4% |
| Functional Employment of Engineers: | | |
| Product Development | 52% | 55% |
| Manufacturing | 48 | 45 |
| Percent Engineers on Board of Directors | 50+% | 22% |

*Source*: L. Harbeck, "Technical Manpower Characteristics of the U.S. and Japanese Automotive Industries," Research Report, Ann Arbor, Mich., Joint University Automotive Study, 1983.

Within the Japanese automobile industry, 7.8 percent of those employed are engineers and 6.2 percent are technical assistants; these two professions comprise approximately 14 percent of the total number of employees. The corresponding figures for the U.S. automobile industry are 4.1 percent and 2.6 percent, respectively, totaling only 6.7 percent of all employees. Therefore, the automotive work force in Japan has a significantly higher

[21] Ibid.

[22] L. Harbeck, "Technical Manpower Characteristics of the U.S. and Japanese Automotive Industries," Research Report, Ann Arbor, Mich., Joint University Automotive Study, 1983; and F. Kodama , Y. Yakushiji, and M. Hanaeda, "Structural Characteristics of the Japanese Automotive Supplier Industry," Working Paper Series No. 13, Center for Japanese Studies, University of Michigan, June 1983.

percentage of professional engineers than it does in the United States. In terms of the functional employment of the engineering graduate, almost half of the degree-holding engineers are found in manufacturing, with the Japanese percentage a bit higher than the U.S. percentage. It is often said that this more pronounced orientation of graduate engineers toward manufacturing in general gives Japan a competitive advantage, especially in this age of the high technology in which product development and manufacturing are becoming more tightly coupled. However, it is to be seen whether this tendency will continue. Perhaps younger Japanese engineering graduates are more inclined to be assigned to the research and development and product design jobs than to manufacturing.

This table reveals also that engineers are found on Japanese boards of directors twice as often as in the United States. This higher degree in representation of engineers is also giving a competitive edge to Japan in the high-technology age, in which corporate management has to be consolidated with research and development.[23]

However, the total number of engineering graduates employed by the automotive industry is much larger than for the graduates of other disciplines such as humanities and social sciences. Therefore, a little over 50 percent of degree engineers on a board of directors does not necessarily mean that the chance for a degree engineer to be promoted to the corporate manager's level is higher than the chance for a university graduate from another discipline. In fact, the chance for a degree engineer is less than that for a non-engineering university graduate, as shown in Table 6-2.

*Table 6-2.* Graduates Employed by Transport Equipment Manufacturing, by Disciplines

(1988)

| | Humanity/Social Sciences | Science/ Engineering | Sub-Total | Grand-Total |
|---|---|---|---|---|
| Transport Equipment Manufacturing | 1,623 (28%) | 4,098 (72%) | 5721 (100%) | 6047 |
| All Manufacturing | 29,015 (45%) | 35,912 (55%) | 64927 (100%) | 75143 |

*Source*: same as Table 1-1
*Note*: Employment statistics are available only under the category of transport equipment manufacturing, of which the automotive manufacturing comprises a majority.

[23] F. Kodama, "Research and Development Dynamics of High-Tech Industry: Toward the Definition of High Technology," *Journal of Science Policy and Research Management* 1 (1986): 65-74; F. Kodama, "High-tech Developments Render Traditional Management Obsolete, *The Japan Economic Journal* 26 (October 1988); and F. Kodama, "Technological Innovation Drives Structural Changes in Corporations," *The Japan Economic Journal* 26 (June 1988).

Science and engineering bachelor's graduates comprise almost three-quarters of the subtotal, which includes humanity and social-sciences bachelor's graduates as well. This number is far larger than 50 percent, the share of degree engineers on a board of directors. Should this reality be recognized widely, an able and ambitious young Japanese people might no longer be motivated to study engineering.

## Conclusion

Japan must be careful in making any drastic policy decision that might alter the existing structure of its university system. Without a detailed study and a deeper understanding of the problems, it is difficult and sometimes dangerous to implement drastic changes that might affect the whole structure of the engineering educational system, which has been developed through a long history of trial and error. Clearly Japan's demand for engineers is growing. What is not so clear is how that demand will be met. We need more systematic investigations of the issues concerning engineering education.[24]

---

[24] F. Niwa, "The Japanese S&T Indicator System and Industrial R&D Resource Diversification," presented at the NISTEP International Conference on Science and Technology Policy Research: "What Should Be Done? What Can Be Done?" Izu-Shimoda, Japan, February 2-4, 1990.

# EDUCATING AND TRAINING THE U.S. WORK FORCE FOR THE TWENTY-FIRST CENTURY

HARVEY BROOKS
*John F. Kennedy School of Government*
*Harvard University*
*Cambridge, MA 02138 USA*

## Introduction

The main arguments of this paper can be summarized in the following propositions:

1. U.S. national competitiveness – and particularly its future competitiveness – will depend more on the technical literacy of its labor force than on the number of its scientists and engineers.
2. The average K-12 education is increasingly mismatched to the requirements of the future work force, even if there were no change in the demographic composition of that work force.
3. In the future new entrants to the labor force will be drawn increasingly from groups that have been poorly served by the public K-12 system and as a result are weak and uninterested in mathematics and science.
4. U.S. economic performance is deteriorating so rapidly that we cannot afford to simply wait for educational reform to take effect, important as that is. We must mount a major national effort to upgrade the skills of the currently employed work force – not so much specific work-related skills as the basic language and math skills necessary to learn the new skills required by technology and market structure change.
5. The technologies in the workplaces of the future will require managers to give low-level workers much greater discretion and responsibility. To continue improving quality and productivity will mean drawing on the talents and experience of workers at all levels, not just scientists and engineers. The most competitive organizations will be those that can foster continuous collective learning from the bottom to the top of the organization.

## Rising Concerns about the Quality of the Labor Force

It is by now conventional wisdom that the public precollege education system of the United States is not adequately preparing the work force it will need in the 21st century.



*D. S. Zinberg (ed.), The Changing University*, 129–140.

At the precollege level there is a consensus that the major problem is the quality of the students being produced. But at the postsecondary level the consensus is that quality is adequate, but the problem is quantity – largely the result of the inadequate quality of preparation at the precollege level, which reduces the potential pool. There is also some concern that engineers, especially in the more elite institutions, are too oriented toward careers in research and development relative to design, manufacturing and applications engineering – the "downstream" segments of the technical enterprise.

There is a growing conviction, however, that adequate numbers of competent scientists and engineers are not enough by themselves to ensure international competitiveness, that the threat to U.S. industrial leadership arises more from the declining quality of the workers that do not go beyond high school (and frequently do not graduate from high school) than from the quality of its scientists and engineers.[1] This problem the U.S. faces concerns most of the population, the millions who will have to operate, supervise, and maintain the increasingly complex technological equipment on whose reliable, safe and predictable functioning everyone depends – the airline pilots and mechanics, the tanker captains, the safety inspectors, the construction engineers, the medical technicians, the nuclear reactor operators, the skilled factory workers and chemical plant operators, the environmental, health and safety regulators.

In addition there is a growing consensus that one of the principal competitive weaknesses of the U.S., particularly compared with Japan, is its failure to integrate research and development, product design, manufacturing and marketing in the innovation or new-product development process. The consequence is that U.S. companies often lose an initial technological advantage because they are slower to introduce new products. Japanese companies introduce new products in two-thirds the time it takes U.S. companies.[2] Accelerating the product cycle means that marketing and other nontechnical people must be able to communicate easily with engineers. This requires a common base of technical literacy among all the people involved in the process, as well as communications and interpersonal skills on the part of the technical people.[3]

To be sure, there is much debate not only in the United States but in most other industrial countries as to what constitutes scientific literacy, either for work or for intelligent citizenship, and how much exposure to mathematics, science and technology is essential for a capable work force.[4] Cyert and Mowery, for example, concluded, "The

---

[1] Harvey Brooks, "The Technological Factor in U.S. Competitiveness," *Business in the Contemporary World*, 2 (Autumn 1989): 81-86.

[2] Kim B. Clark, W. Bruce Chew, and Takahiro Fujimoto, "Product Development in the World Auto Industry," *Brookings Papers on Economic Activity* 3 (1987): 729-781.

[3] For a rather dramatic account of one attempt to speed up the product development cycle and the skills involved, see N. R. Kleinfield, "How 'Strykeforce' Beat the Clock: Ingersoll-Rand Did the Impossible: Collapsed the Design Cycle to One Year," *New York Times*, Sunday, March 25, 1990 (Business Section).

[4] Cf., for example, Herbert J. Walberg, "Scientific Literacy and Economic Productivity in International Perspective," *Daedalus: Scientific Literacy* (Spring 1983): 1-28.

formal training system [comprising elementary, secondary and higher education] for labor market entrants [16 to 24 years old]...appears to respond rapidly to changes in the skills demanded by employers."[5]

Other observers, however, point to a growing mismatch between the quality of the work force and the mix of skill requirements of the new jobs being created in the economy.[6] Growth in unemployment, particularly long-term unemployment, among the unskilled and the undereducated is only one symptom of this mismatch. All of this is in spite of an economy that has an unequaled performance in creating new jobs during the past 15 years.[7] Moreover, after nearly a decade of economic expansion and employment growth, the gap in living standards between the worst and best educated is widening in the United States, reversing a long-term trend toward greater economic equality that started during World War II.

Concern about the quality of the work force, particularly those without postsecondary education, is by no means confined to the U.S., although it is certainly the most acute there. According to one commentator, this concern "gnaws at the heart of every advanced industrial nation. The more complex the economy and the more heterogeneous the population, the greater the problem. It is a central factor determining the national standings of the global economic competition, and there is not a single industrial nation that is not dissatisfied with the quality of the products of its education and worker training systems."[8]

Other industrialized countries besides the United States have experienced problems in assimilating educationally disadvantaged minorities into their work forces and hence into the mainstream of their societies. Most of the countries of Western Europe have a significant population of the descendants of guest workers imported during the postwar recovery period, who have still not been fully assimilated into the educational system or mainstream work force.

## Matching the New Technology and the New Work Force

With the rapid introduction of information technology, any job that can be specified in writing by engineers or planners can and will be automated out of existence. Computers

---

[5] Richard M. Cyert and David C. Mowery, eds., *Technology and Employment: Innovation and Growth in the U.S. Economy* (Washington D.C.: National Academy Press, 1988), p. 142.

[6] William B. Johnston and Arnold H. Packer, *Workforce 2000: Work and Workers in the Twenty-First Century*, HI-3796-RR (Indianapolis, IN: Hudson Institute, June, 1987).

[7] Cyert and Mowery, op cit., p. 53. The U.S. economy created 20.7 million new jobs in the decade 1975-85, an average of more than 2 million a year. In most of Europe, employment shrank slightly during this period, and the growth in Japan was much smaller than in the United States.

[8] William E. Nothdurft, *Schoolworks: Reinventing Public Schools to Create the Workforce of the Future* (Washington, D.C.: Brookings Institution, 1989).

can be taught to execute such jobs with a much lower error rate than human beings, who suffer from inevitable lapses of attention.[9] The newer microelectronics-based technologies, if they are to be effective in getting the job done better, require an organization of work that depends on discretion and independent judgment on the part of ordinary workers on the shopfloor and in the office. It is no longer sufficient for such workers to simply take orders without question from supervisors and experts; they must be able to make decisions based on their personal understanding of the overall purposes and strategies of the organization for which they are working.[10]

In the next 15 years only 15 percent of the new entrants to the U.S. labor force will be white males, as compared with 47 percent white males in the current labor force. The labor force will be increasingly made up of people whom the U.S. public education system has poorly served and who have not traditionally been well represented in mathematics and science programs[11] nor in technical occupations and professions. A similar situation will exist in other mature industrialized countries, as their labor forces become increasingly populated with second- and third-generation descendants of immigrant guest workers whose birth rates have considerably exceeded those of the indigenous population.

Many other industrial countries have much better-developed public vocational-training systems that reach a larger fraction of the non-college-bound population and provide both training for specific jobs and the basic cognitive skills necessary for quickly learning new jobs in the future. But even the countries with better job-training systems than the United States are beginning to worry about the adequacy of their systems for the future, especially in the light of their increasingly culturally heterogeneous populations.

In all the industrialized economies, competitive advantage rests increasingly on the quality of their human resources; traditionally such advantages as transportation routes, natural resources and climate played a greater role. Technological prowess has become an important measure of international political standing and influence as well as of economic capacity. This is perhaps more so than is justified by the realities of global interdependence, in which even the most capable nation can no longer hope to be technologically self-reliant.[12]

---

[9] Robert U. Ayres, "Complexity, Reliability, and Design: Manufacturing Implications," *Manufacturing Review* 1 (March 1988): 26-35.

[10] Carlota Perez, *Technical Change, Competitive Restructuring, and Institutional Reform in Developing Countries* (Strategic Planning and Review Discussion Paper No. 4, The World Bank, Strategic Planning and Review Department, December, 1989); see also Harvey Brooks and Michael Maccoby, "Corporations and the Work Force," Chapter 6, pp. 113-132, in John R. Meyer and James M. Gustafson, eds., *The U.S. Business Corporation: An Institution in Transition*, published for the American Academy of Arts and Sciences (Cambridge, MA: Ballinger Publishing Company, 1988).

[11] Nothdurft, op. cit.

[12] Harvey Brooks, "National Rivalries and International Science and Technology," in Karl Vak, ed., *Complexities of the Human Environment: A Cultural and Technological Perspective* (Europa Verlag GsmbH Wien, 1988) pp. 49-62.

Not everyone agrees that the industrial economies of the future will require higher average skill levels than in the past. Concern about the shortages of skilled people still coexists with concern about the displacement of skilled industrial workers by automation, a concern that has been partly triggered by the disappearance of high-paying manufacturing jobs in heavy industries of the "rustbelt," not only in the United States, but also in some of the industrial heartlands of Europe and even Japan.

Some believe that the increased skill requirements associated with information technology (IT) today is only a temporary phenomenon brought about by the relative immaturity of the technologies. High skill levels are frequently needed when new technologies are first introduced, but they soon become more "transparent" and "user friendly" as well as reliable. It is only a question of time before troubleshooting becomes more systematized and routine, requiring less skill and imagination. As more and more intelligence is incorporated into hardware and software, less and less equipment-specific training will be needed. Indeed, some observers have suggested that the puzzling failure of information technology to dramatically boost productivity is a consequence of its immaturity and its slow assimilation into work organizations.[13] According to this view, innovations come in closely interrelated clusters based on a new technological paradigm, thereby producing extra demands on human skills, which will gradually disappear as the new paradigm becomes fully assimilated into the production organization.

Systematic on-the-job training of the work force is an important part of this assimilation process. But there is also a danger that in the current enthusiasm for technical literacy, too much emphasis will be given to equipment-specific training, while the technology is still rapidly evolving. The requirements for working in a technological society are more subtle and difficult to define than can be encompassed by any repertoire of job-determined skills. They cannot be achieved by filling people's heads with scientific facts or by teaching modern science's latest findings or by explaining the workings of a single technological system.

## Scientists and Engineers

The nature of modern economic activity – including both the demand for technological innovation and public demand for the regulation of technology – seems to ensure that the supply of scientists and engineers will be one of the principal pacing factors in economic growth and competitive performance in the next several decades. Demand for scientists and engineers is expected to grow considerably faster than for other types of workers, unless the release of technical people from military work occurs faster than is now

---

[13] Paul A. David, "General Purpose Engines, Investment, and Productivity Growth: From the Dynamo Revolution to the Computer Revolution," Center for Economic Policy Research, Stanford University, prepared for Royal Swedish Academy of Sciences Symposium on "Technological Change and Investment," Stockholm, January 21-24, 1990.

anticipated.

There is actually a close connection between the work force's inadequacy for future job opportunities and the lack of scientists and engineers. If the United States solves the problem of producing a future population of scientifically literate workers in the 21st century, this population will provide an adequate pool from which to draw future scientists and engineers. If it does not, there is likely to be a serious shortage, and the country will continue to increase its dependence on immigrants for many positions requiring a high level of technical training. At present this dependence is probably a net benefit, because the foreign students who come and remain to work in the United States tend to be the most talented representatives of very large populations – a much more highly selected group than native-born students. However, many of these individuals, having broadened their experience and sharpened their skills by working in U.S. laboratories and industries, will increasingly choose to return to their own countries as their economies take off and become more technology intensive. This is already happening with South Korea and Taiwan.[14] This reverse flow will be accelerating at just the time when new entries to the United States' technical work force will come more and more from the educationally disadvantaged groups in the U.S. population.

## Feasibility of the Goal of Universal Technical Literacy

One of the principal objections often posed to the goal of universal scientific literacy is the enormous variation in ability of the population. Surprisingly enough, this objection is much more prevalent in the United States than it is in many countries that have a much more pronounced class structure. The curriculum-reform movement in science and mathematics in the United States that followed the launch of Sputnik in the early 1960s and the resulting space race, for example, was designed to attract the most able students into science and engineering. The current wave of curriculum reform, however, is based on the goal of everybody achieving a minimum level of science and math literacy that was implicitly assumed to be possible only for a small minority two decades ago. A recent report of the Mathematical Sciences Board of the National Research Council takes the position that all students can learn mathematics, pointing out that the United States is almost the only advanced country where it is still assumed that learning mathematics requires a special talent found in only a small fraction of the population. According to this report, "one of the most disturbing conclusions of recent studies of mathematics education is that the American public tends to assume that differences in accomplishment in school mathematics are due primarily to differences in innate ability rather than

---

[14] Stephen K. Yoder, "Reverse 'Brain Drain' Helps Asia but Robs U.S. of Scarce Talent," *Wall Street Journal*, April 18, 1989.

differences in individual effort or in opportunity to learn.... [thus] public acceptance of deficient standards contributes significantly to poor performance in mathematics education."[15]

While there is little doubt that there are large differences in innate ability, it is less clear whether these differences actually affect *what* people can learn or only the amount of time and effort it takes to learn it. Simon and his colleagues have evidence indicating that the same fundamental thought processes are involved in elementary and advanced science.[16] "The development of literacy, the acquisition of information, and the problem solving of beginners differ in degree rather than in kind from the mental activities of experts. The scarce resources are opportunity and concentration rather than the amount of information available or the processing capacity of the mind both of which are, for practical purposes, unlimited," Walberg states.[17] One researcher has demonstrated that an undergraduate of average intelligence before training could remember only 7 digits, but could be trained to remember 79 digits. That is almost 10 times more than the number of digits that an individual rated by psychologists as of superior intelligence can be expected to remember without special training.[18]

If such simple experiments as this can be considered a paradigm for higher skills as well, the implications for education could be enormous. They suggest that training could offset initial differences in innate ability regardless of whether these are of genetic or cultural origin. Realizing such a possibility for higher skills will almost certainly require the results of extensive research in cognition, which we lack at present; the experiment cited above required 230 hours of instruction and practice based on cognitive science. Of course, it might take more than the human life span to turn a dullard into a genius, but more modest ambitions of universal scientific literacy aspired to by recent initiatives in curriculum development might still be achievable.[19] In the era of Fordist mass production, which operated on a theory that geniuses could design systems that could be operated by morons, one might have argued that it was economically efficient to invest most educational effort in the people who were most educable, that is, who could be advanced the furthest with a given investment of resources – or, still in other words, the people of highest "innate ability." But as higher levels of education become necessary for a larger proportion of the work force, such a strategy may no longer be the most cost-effective. As routinely used technologies become more complex and expensive, the costs of an undereducated work force have become so high that a greater and more efficient

---

15 *Everybody Counts: A Report to the Nation on the Future of Mathematics Education* (Washington, D.C.: National Academy Press, 1989).

16 Herbert A. Simon, *The Sciences of the Artificial* (Cambridge, MA: MIT Press, 1981).

17 Walberg, op. cit., p. 4.

18 K. A. Ericsson, W. G. Chase, and S. Faloon, "Acquisition of Memory Skill," *Science*, 208 (1980): 1181-1182 (quoted in Walberg, op. cit.).

19 National Council on Science and Technology Education, *Science for All Americans* (Washington, D.C.: American Association for the Advancement of Science, 1989).

investment in a larger and larger fraction of the work force has become essential. A human-resource investment too concentrated on the top end of the innate-ability distribution may no longer be well matched to investment in technology. That is apparently the rationale behind the current educational reform movement, even though it has been articulated more in terms of equity than efficiency.

Yet, today both the formal educational system and the systems for training employees in industry are still operating on the principle that at each stage you should invest the most in those who reach that stage having had the greatest previous educational investment. There is ample evidence that job-related training is much more available to individuals who already have a high level of education; the higher the educational level, the greater the average investment by the employer in training. In the words of one report:

> Although worker retraining has become a catchphrase, and the federal government and private industry now spend billions of dollars for retraining, there is still no national consensus that all workers should expect to learn new skills over the course of their work lives. Except in a few companies, training is confined mostly to the top and bottom ranks of employees, with little systematic effort to ensure that all workers are constantly reinvesting in themselves to avoid obsolescence.[20]

Further evidence of the feasibility of much more universal literacy, both regular and technical, comes from the experience of the Japanese. According to Walberg, during the postwar-period average IQ in Japan rose from 103 to 111.[21] Today 77 percent of the Japanese population has an IQ greater than 100, the typical median for Western countries, which has not changed since the end of World War II.[22] Walberg points out that "since heredity cannot determine large changes [of IQ] over short time periods, the change is most certainly attributable to social environment, family circumstances, education, health, and other human capital investments and enhancements."[23] Given the intensity of Japanese precollege education and the percentage of the curriculum devoted to science and math, one might conclude that education was the major contributor to this remarkable performance, which is consistent with the experience of a few U.S. experimental schools.[24]

---

[20] Johnston and Packer, op. cit., p. xxv.

[21] Walberg, op. cit., p. 21.

[22] Richard Lynn, "IQ in Japan and the United States shows a Growing Disparity," *Nature*, 197 (May 1982): 222-223.

[23] Walberg, op. cit., p. 22.

[24] Ibid., p. 18.

## Integration of Education with Work

In the United States, where job mobility is high, there is a disincentive for employers to provide training that may end up benefiting a competitor. It is true that the United States does have a public vocational training system, but it has a poor reputation and does not appear to meet training needs as seen by employers. Many other industrial countries do provide higher levels of job-related skills training to blue-collar workers than does the United States, in part because such training is seen as a public good and therefore its cost is shared by the public or provided publicly in close cooperation with industry. That is reflected, some say, in more rapid adoption of new manufacturing technologies in these countries than in the United States.[25]

The countries that are strongest in employment-related training are those whose national technology strategies Henry Ergas characterizes as diffusion oriented.[26] Examples are West Germany, Switzerland, and Sweden, where much more effort is made to integrate the world of education and the world of work than in the United States.[27] This is especially true for workers without higher education. The oldest and most extensive system is that in the Federal Republic of Germany, which goes back more than 100 years. It has the following characteristics:

1. Industry supports and significantly funds the program.
2. It involves both theoretical training and hands-on apprenticeship training in an industrial setting. The mentorship responsibility on the job is taken very seriously within companies.
3. A high percentage of all school leavers benefit from the system, which is one reason that youth unemployment in Germany is no higher than general unemployment, a situation completely different from that in the United States, Britain and many other industrial countries.
4. For fragmented industries such as machine tools and other metalworking, industry associations play a big role in funding and designing the training.
5. The training system is directed toward the rapid diffusion of technological advances in relatively mature, traditional industries and is strongly craft oriented.

In Sweden the training for non-college-bound youths is more school centered and is aimed at the smoothest possible transition from formal education to work, with a good deal of individualization, as part of a social contract to ensure employment opportunities

---

[25] R.R. Nelson, M.J. Peck, and E.D. Kalachek, *Technology, Economic Growth and Public Policy* (Washington, D.C.: Brookings Institution, 1967).

[26] Henry Ergas, "Does Technology Policy Matter?" pp. 191-245 in B.R. Guile and H. Brooks, eds., *Technology and Global Industry: Companies and Nations in the World Economy* (Washington, D.C.: National Academy Press, 1987).

[27] Nothdurft, op. cit.

for all. Rotation into work settings starts quite early in schooling, and schooling continues into working life. Sweden, besides having one of the largest R&D/GDP ratios of any industrialized country, spends a comparable fraction of its gross domestic product on public labor-market activities. These expenditures include training and retraining as well as an elaborate computerized system for matching job seekers to job opportunities. As a result, Sweden has had one of the lowest unemployment rates among industrial countries.

Both Sweden and Japan have been exceptionally successful in phasing down sunset industries such as shipbuilding and steel without large spurts of unemployment. That can largely be attributed to the excellent training of their workers – in Sweden mostly by the public sector, in Japan mostly by the private sector through the tradition of guaranteed employment in the large companies. U.S. companies with full-employment policies, such as IBM, AT&T, and Hewlett Packard, are also noteworthy for their high investment in employee training. However, increased global competition may make employment guarantees impractical for individual companies; hence, it is likely that more responsibility will be shifted to the public sector in all countries, since human resources with polyvalent skills are becoming more and more a public good.

Even countries with the best special-training and apprenticeship systems for craftsmen and technicians, such as Germany, are beginning to anticipate that new generations of production technology will succeed each other so rapidly that specialized skills learned primarily on the job will become obsolete several times within the working life span of the typical worker. Moreover, a system designed to ensure the rapid spread of incremental improvements frequently runs into difficulties if there is a fundamental discontinuity in the technological paradigm, as happened with the advent of computer-controlled machine tools in Germany and digital watches in Switzerland, although both countries managed to recover their positions through additional training in electronics before it was too late. This is leading these countries to rethink the proper mix between school training and practical skill training on the job. Germany is considering increasing the time devoted to theoretical training and further development of conceptual skills in apprenticeship training periods.[28] The dual track of apprenticeship and academics is also feared to be losing its effectiveness because the apprenticeship systems are not keeping up with changes in production technology.

## The Importance of Research

One of the characteristics of modern industrial organizations is that they must be learning organizations. Their watchwords are "continuous improvement" and "dynamic competition." The global competition of recent decades has taught us that nothing fails like success; today's winning formula is tomorrow's disastrous strategy. A successful competitor is almost never in a steady state; the more successful it is, the more quickly

---

[28] Ibid.

it will attract imitators and see its market saturated. Unfortunately, schools are about as far removed from being learning organizations as it is possible to get. They have few ways of knowing how they are doing. Successful experiments diffuse slowly and are often forgotten. Teaching is still regarded as a "craft," and little of practical applicability is known about how people learn or how they use what they learn.

In 1987 about $160 billion was spent on elementary and secondary education in the United States, and less than 0.1 percent of that was spent on educational research and data collection.[29] Very little has been spent on what would be called fundamental research in other contexts – studies of the phenomenology of learning – and little that has been learned from cognitive science has been applied to the learning process, although according to many experts, a good deal of what has been learned in the past few years should be fairly immediately applicable – for example, "that learning is not 'filling up an empty glass' but requires emptying out some wrong ideas; that true understanding of some concepts is difficult; that successful education starts from where the learner is."[30] Good teaching requires a much more sensitive understanding on the part of the teacher of just where the learner is in his or her cognitive development.

In a recent report, the Committee on Economic Development, a business research organization, had this to say about the low levels of educational research spending in this country:

> Private industry could not succeed with a data collection system and research base as weak as this nation has in the field of education. Yet it is only through education research and data collection that we can expect to identify ways and means to increase the output of the education system. The original purpose of the federal role in education was as a national repository of information, and the first federal Department of Education, launched in the 1860s, was designed to accomplish that objective.[31]

Educational research becomes much more critical when the emphasis is on bringing the entire population to a higher minimum level of mathematical and scientific literacy. That is how the current reform strategy differs from that following the Sputnik crisis of 1957, where the emphasis was on identifying and training an academic elite, although this assumption was probably never fully understood by the movement's proponents at that time. Thus, if the current reform impulse is to achieve its ambitious goal of conveying to all Americans a "common core of learning... limited to the ideas and skills that have the

---

[29] David T. Kearns and Denis P. Doyle, *Winning the Brain Race: A Bold Plan to Make Our Schools Competitive* (San Francisco, CA: ICS Press, 1988) p. 106.

[30] David Z. Robinson, "Fight in the Corner Where You Are: Starting to Improve Pre-College Education," Keynote Speech, American Medical Association, National Initiative for Science and Technology Education, March 14, 1989.

[31] *Investing in Our Children: Business and the Public Schools*, Committee on Economic Development (CED), 1987.

greatest educational and scientific significance,"[32] a much deeper theoretical understanding of the learning process will be required, including an understanding of the diversity of cognitive styles and of the different educational strategies they may require. The problem is made much more complicated because the educational task is embedded in an increasingly difficult cluster of social and motivational problems. Can techniques derived from deeper knowledge of cognitive science overcome all the obstacles created by the social and family environments in which so many children grow up?

Achievement of the goal of scientific literacy for all will almost certainly require breaking the age lockstep – the concept that all students move through the educational system at about the same rate, regardless of innate ability or cognitive style. This is especially true at the lower levels. How can this be done without destroying the motivation of the slower students? To what extent can the imaginative adaptation and use of modern information technology overcome this problem? Ever since computers first came into widespread use in the 1960s, experts have predicted an imminent revolution in educational technology, but this revolution has receded ever further into the future. Some progress has been made, but least in the formal educational systems. To what extent is this disappointment due to the stubborn resistance of a craft-oriented profession, and to what extent to the inadequacies of the technology and the failure to adapt it to the social, economic and psychological realities of the real learning situation? On this there is little consensus.

Fundamental research on learning is likely to have its earliest and largest payoff not in the schools but in the workplace, where the environment is already hospitable to adopting the results of research and to investing in technology. However, without carefully crafted public support, the incentives in the private sector are likely to continue to concentrate educational resources on the most easily educable.

U.S. education, particularly at the lower levels, may give insufficient emphasis to the development of systematic work habits and self-discipline for fear of dampening the originality and individual creativity that are prized so highly in U.S. culture. This is also an important subject for cognitive research. In a recent book based on observations of schools in modern China, Howard Gardner of the Harvard Graduate School of Education has raised this issue, suggesting that some mixture of Chinese discipline and the U.S. stress on creativity and spontaneity may be better than either one alone.[33]

---

[32] National Council on Science and Technology Education, op. cit.

[33] Howard Gardner, *To Open Minds: Chinese Clues to the Dilemma of Contemporary Education*, (New York: Basic Books, 1989).

# THE CHANGING PATTERNS OF INTERNATIONAL COLLABORATION IN UNIVERSITIES

JEAN-FRANÇOIS MIQUEL[1]
*Office of Technology Assessment*
*United States Congress*
*Washington, DC 20510-8025*

## Abstract

International cooperation in the academic sector is becoming more frequent and therefore plays a growing role in the production of scientific knowledge. This trend is demonstrated by international coauthorships in research literature. The purpose of this study is to measure the internationalization of science through the scientific relationships among universities. The authors present a matrix constructed from coauthorship data and introduce indicators. Using these measures, they observe two scientific relationships – between the United States and Greece and between the United States and Denmark. Such examples should reveal different cooperation patterns between countries and lead to a better understanding of scientific communication patterns in the mainstream of science.

## 1. Introduction

Every scientist would like to work with another scientist who could contribute to his or her research. Scientists know each other's work through various networks: articles, seminars, conferences, among others. Thus when they meet, for example, at an international conference, they may not only exchange information, but may also plan to initiate a collaborative project. Cooperation may start by the exchange of doctoral and postdoctoral fellows between laboratories. And through facsimile machine and electronic mail, they may exchange data on everyday experiments. Later they may coauthor their results in an article. This kind of story repeats itself at an increasing rate.

Many collaborative projects, thus started, occur outside formal intergovernmental agreements or programs. Informal cooperation is often not recorded in financing agencies, academies, or research councils, because negotiations for these projects are conducted between the universities and laboratories. Cooperation could be traced only by the result of each collaborative work, which most often takes the form of coauthorship. The end product of scientific research always finds expression in scientific literature, and the

[1] On leave from Laboratoire d'Evaluation et de Prospective Internationales, Centre National de la Recherche Scientifique, 295, rue Saint-Jacques, 75005 Paris, France.



*D. S. Zinberg (ed.), The Changing University*, 141–151.

essential part of this literature is found in scientific journals.

Through the *Science Citation Index*, it is possible to identify the collaborating countries and the affiliations of research institutions that conduct international research.

## 2. Methods

Coauthorship is defined in this study as international institutional coauthored articles.[2] The coauthorship data derives from International Science Indicator's *Science Citation Index*, compiled by CHI.

This data contains the number of internationally coauthored articles, notes, and reviews in more than 3,000 journals published during the period 1981-86.

Coauthorship data gives two indicators: COA (Coauthorship) and COP (Cooperative effort).

Suppose that there is a coauthorship (COA) between the U.S., Greece and Denmark. If we focus on the United States' international activity, through this collaborative project the United States creates one link (COP) with Greece and one with Denmark (thus two links altogether) and produces one COA in the international scientific community.

The COA indicator gives the amount of output issued from cooperation, thus measures productivity.

The COP indicator measures the cooperative efforts made by countries through writing articles and the number of scientific links created between countries through common work. Thus, COP measures the degree of the scientific ties between countries and the degree of participation of a country in the international scientific community.

COA and COP are different measures, but complement each other. Both indicate trends in international relationships between countries.

Based on this concept, a matrix of 72 countries has been created in eight fields of science, for each year from 1981 to 1986 (MEV-MAC Matrix). This matrix enables the observation of relationships between countries from various angles.

In this study, the "entire world" (WRD) means the 72 countries in the MEV-MAC Matrix.

The eight fields of science are the following: mathematics (MAT), physics (PHY), chemistry (CHM), engineering and technology (ENT), earth and space sciences (EAS), biology (BIO), biomedicine (BIM), and clinical medicine (CLI).[3]

---

[2] *Science and Engineering Indicators* (Washington D.C.: National Science Board, 1989) and R.J.W. Tijssen and H.F. Moed, *Science and Technology Indicators* (Leiden: DSWO Press, 1989).

[3] M.P. Carpenter, "International Science Indicators -- Development of Indicators of International Scientific Activity Using the *Science Citation Index*," Report for NSF contract SRS77-22770 (1970).

## 3. International Scientific Activities

### *3.1 International Scientific Activity of the United States*

The United States is the largest producer of science, and U.S. researchers participate in from 17% in chemistry to 29.1% in mathematics of international projects of the world, and this indicates the largest participation in the international scientific community among the 72 countries. The number of countries with which it collaborates is large, varying from 59 countries in mathematics to 71 countries in clinical medicine. Thus, the United States is the major collaborating country in international research. The scientific excellence of this country attracts scientists from all over the world.

Two countries of the European Community, Denmark and Greece, have been chosen to observe the different cooperation patterns with the United States.

### *3.2 International Activities of Denmark and Greece*

The scientific sizes of Denmark and Greece, measured by publications, differ. Denmark is approximately three-and-a-half times larger than Greece in scientific production and also in its international scientific activity. Denmark ranks as the 18th largest producer among the 72 countries, whereas Greece is 32nd. Though these two countries differ in scientific size, the proportion that international coauthorship takes in each country's entire scientific publication is practically the same: 23.5% for Denmark and 23.4% for Greece, which indicates that for both countries, about one out of four publications issues from international collaboration. The growth of collaboration between the United States and each of these two countries during the period 1981-86 is a 22.5% increase for Denmark, and a 97.1% increase for Greece although numbers differ in absolute term.

### *3.3 International Collaboration Between the United States and Denmark*

For Denmark, the United States is one of the largest collaborative partners in almost every field of science. Exceptions are in biology, in which the U.K. is a larger partner, and clinical medicine, in which Sweden is the primary collaborator with Denmark. From 1981 to 1986, 1,491 COPs were created between the United States and Denmark, in eight fields of science. This indicates that 21.9% of the entire international activity of Denmark was conducted with the United States for that period.

Figure 1 shows the trends in the cooperation of Denmark with the United States, in five fields of science. In most of the fields the amount of collaboration remained fairly stable, but not in biomedicine and engineering and technology. In biomedicine, the amount of collaboration more than doubled (31 COPs in 1981 to 65 COPs in 1986). The amount of collaboration in engineering and technology was small in 1981 (four COPs), but increased to 18 COPs by the end of the period.

### *3.4 International Cooperation between the United States and Greece*

For Greece, the United States is the largest collaborative partner in every field of science except earth and space sciences, in which the U.K. is the primary collaborator.

**Figure 1**

**Trends in the Cooperation of Denmark with the United States in Five Fields of Science, 1981-86**

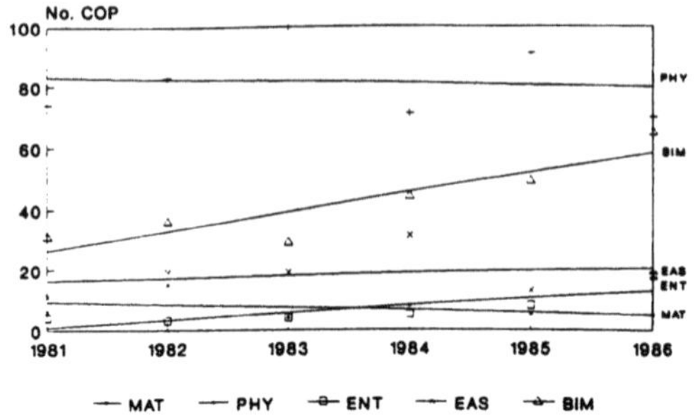


**Figure 2**

**Trends in the Cooperation of Greece with the United States in Five Fields of Science, 1981-86**

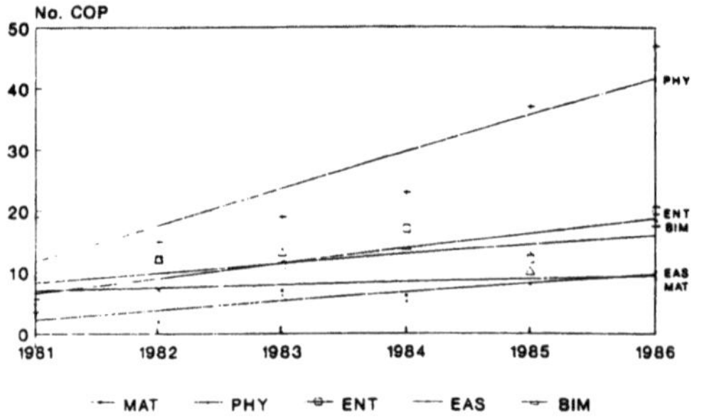


CNRS-LEPI MEV-MAC

Figure 2 shows the trends in the cooperation of Greece with the United States in five fields of science. Greece had 581 COPs in six years with the United States, which indicates that 30.5% of entire international activity of Greece was conducted with the United States. Collaboration increased in almost all fields. Scientific ties between the United States and Greece became stronger, especially in physics, which increased by 147%. In mathematics, engineering and technology, and biomedicine, there was hardly any collaboration in 1981, but in 1986 COPs multiplied.

Figures 3a and 3b contrast the cooperation pattern indicated by percentages of eight fields of science. These figures also indicate the changes in these patterns during 1981-86. The difference of profiles between the two cooperation patterns is greater in 1986 than in 1981. The proportion of life sciences in 1986 is higher in Denmark-United States cooperation (53.0%) than in Greece-United States cooperation (32.3%). Physics, chemistry, and engineering and technology, on the contrary, take a more important proportion in Greece-United States cooperation (53.6%), whereas between Denmark and the United States, these fields account for 38.3%.

**Figure 3a**

**Changes in the Cooperation Pattern of Denmark and the United States: Percentages of Eight Fields of Science**

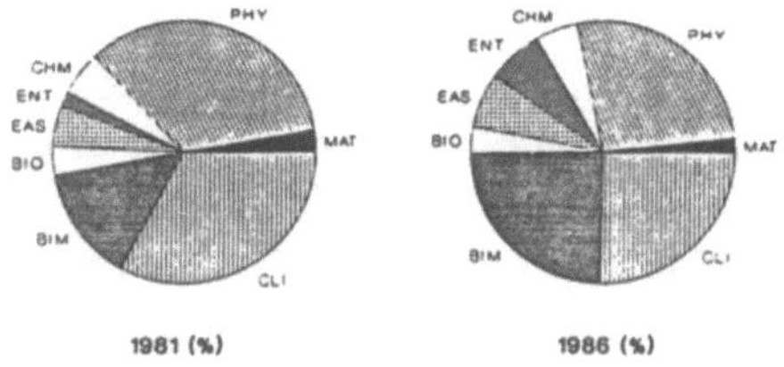


**Figure 3b**

**Changes in the Cooperation Pattern of Greece and the United States: Percentages of Eight Fields of Science**

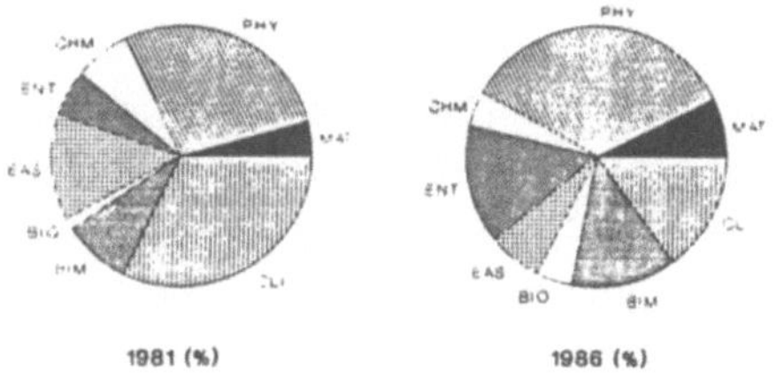


CNRS-LEPI MEV-MAC

## 4. Difference between Denmark-United States and Greece-United States Cooperation Measured by Affinity Index

### *4.1 Affinity Index*

The affinity index (AFI) is a measure of the amount of collaboration between a given country (A) and another (B), compared with the total collaboration of the given country (A) with the entire world, in a given field of science, during a given period of time. AFI is therefore the number of COPs between A and B divided by the total COPs A has with the rest of the world (WRD), in a given field and during a given period of time. It indicates the scientific affinity of A toward B (A→B).

AFI (A→B):

$$\frac{\text{COP }(A\leftrightarrow B)}{\text{COP }(A\leftrightarrow WRD)} \times 100$$

Mutually, AFI for B indicates the scientific affinity of B toward A (B→A).

AFI (B→A):

$$\frac{\text{COP }(A\leftrightarrow B)}{\text{COP }(B\leftrightarrow WRD)} \times 100$$

The affinity index uses the columns of the matrix, therefore, A is a country and B can be a country (e.g., the United States) or a geographical or economic group of countries (e.g., the European Community or the Pacific Rim countries).

The affinity index is used to find how B ranks in A's international activity with the world in a given field. It also reciprocally finds how A ranks in B's international activity with the world in that field.

*4.2 Affinities Toward the United States*

Figure 4 illustrates the affinities of Denmark toward the United States (DNK→USA) and of Greece toward the United States (GRC→USA), in eight fields of science for the period 1981-86. It indicates what proportion collaboration with the United States takes in each country in the eight fields.

As a whole, cooperation with the United States represents a larger proportion of Greece's entire international activity (30.5%) than of Denmark's (21.9%). One of the reasons for this difference is that Denmark has relationships with a larger number of countries than does Greece, 54 partners versus 44; thus, international collaboration is more widely spread in Denmark. In addition, Denmark's relationships with its Scandinavian neighbor countries are especially strong. In fact, the scientific ties between Denmark and the Scandinavian countries together are as strong as those between Denmark and the United States (21.7%), whereas collaboration with northern European countries represents only 4.0% of Greece's entire international activity. This link between Denmark and Scandinavian countries is strongest in clinical medicine, in which 33% of Denmark's entire international cooperation is conducted with another of the Scandinavian countries. However, Greece seeks work largely with U.S. physicians in this field (31.6%). This analysis reflects the strong tradition of training and research of Greek physicians in the United States.

The fields in which cooperation with the United States accounts for the largest proportion of Greece's research are mathematics (48.6%), engineering and technology (46.9%), and biomedicine (40.1%). On the other hand, for Denmark, the important fields are mathematics (30.9%), earth and space sciences (28.9%), and physics (25.3%).

**Figure 4**

**Affinities of Denmark and Greece Toward the United States in Eight Fields of Science, 1981-86**

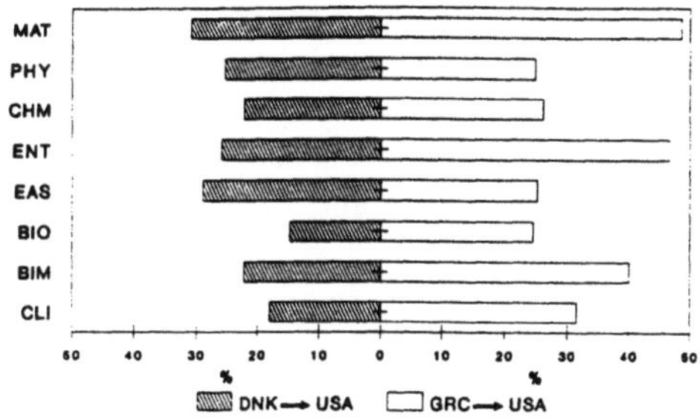


**Figure 5**

**Affinities of the United States Toward Denmark and Greece in Eight Fields of Science, 1981-86**

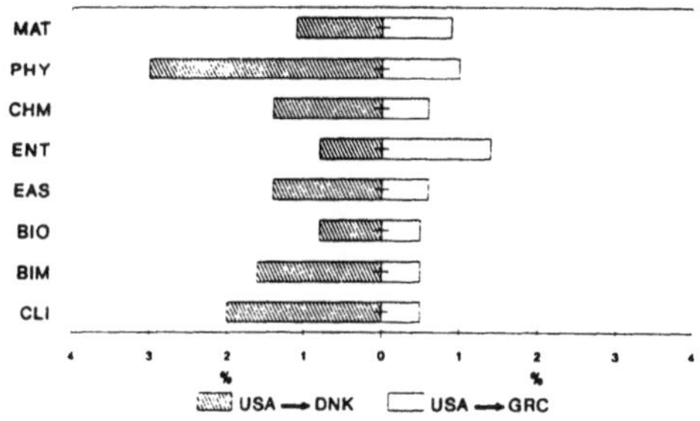


CNRS LEPI MEV-MAC

## *4.3 Affinities of the United States Toward Denmark and Greece*

Figure 5 illustrates the affinities of the United States toward Denmark (USA→DNK) and toward Greece (USA→GRC), in eight fields of science, for the period 1981-86. This figure shows how Denmark and Greece rank in the United States' international activity.

As a whole, Denmark accounts for a larger proportion of the United States' international scientific activity than does Greece. The largest proportion of U.S. affinity toward Denmark is in physics and in clinical medicine: 3% and 2%, respectively. On the other hand, in engineering and technology, Greece has a greater weight of U.S. cooperation (1.4%) than does Denmark (0.8%). This may be a result of a large affinity of Greek scientists toward the United States in engineering and technology.

## 5. Micro-Level Analysis

International cooperation may also be analyzed at the university, laboratory and research institution level or at the level of relationships between researchers.[4]

With the appearance of CD-ROM technology, collaborative projects can be identified more rapidly at this level; it is easier to focus on individual cooperation.

Figure 6 shows international cooperation between Greek universities and universities in California and Massachusetts in eight fields of science in 1986, measured by coauthored papers (COA) between Greek and U.S. scientists. The number of articles authored between Greece and California is practically the same as the number between Greece and Massachusetts, 22 and 20, respectively. However, between Greek universities and Massachusetts universities, coauthorship was concentrated mainly in life sciences, 13 out of 22 COAs. Most collaborative works were conducted between Harvard University and the University of Crete. And nine out of these 13 coauthorships in life sciences were written by the group of Professor Fotis Kafatos.

On the other hand, with Californian universities, collaboration was more active in physics and in engineering and technology than in life sciences, 13 and five, respectively. Most collaborative works in physics were conducted between the University of Southern California and the University of Crete, with the group of Professor Lambropoulos accounting for a large portion of this collaborative research.

These two professors, Kafatos and Lambropoulos, after visiting for many years at U.S. universities, created the Center of Research of Crete (CRC) in Heraklion in 1982.[5] Here we can observe a typical example of initiating collaboration. Relations thus started may continue to bring further cooperation for other scientists.

Figure 7 shows international cooperation between universities in Denmark with universities in California and Massachusetts in eight fields of science in 1986. Collaboration was much more active with Californian universities than with Massachusetts universities, but in both cases life sciences were the major scientific activity. With universities in Massachusetts, 16 out of 21 COAs were in life sciences. Collaboration was undertaken largely with Harvard University (8 COAs) and some with Tufts University (4 COAs). Earth and space sciences and physics were mainly with the Massachusetts Institute of Technology.

Between research institutes in Denmark and those in California, 17 out of 36 COAs were in life sciences. U.S. partners vary in every collaborative project, and unlike the collaboration pattern of Greece, it is difficult to focus activities on particular institutions, as Danish cooperation depends more on individual scientists undertaking collaborative projects spontaneously. However, collaboration in physics was mainly conducted by

---

[4] J.F. Miquel, Y. Okubo, N. Narvaez and L. Frigoletto, "Les scientifiques sont-ils ouverts a la cooperation internationale?," *La Recherche* 20 (1989): 116-118.

[5] D. Markovitsi, "Une Technopole en Crete: Reve ou Realite?," *CNRS Report* (1989).

Nordita and the Niels Bohr Institute in Denmark and the Lawrence Berkeley Laboratory of the University of California at Berkeley in the U.S.

**Figure 6**

**Number of Coauthorships between Greek Universities and Universities in Massachusetts and California in Eight Fields of Science, 1986**

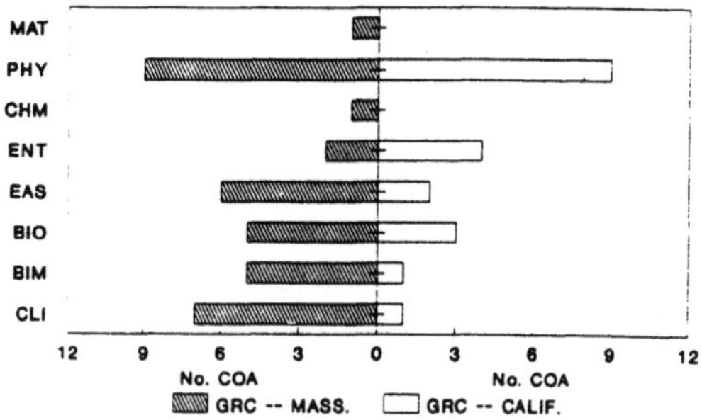


**Figure 7**

**Number of Coauthorships between Danish Universities and Universities in Massachusetts and California**

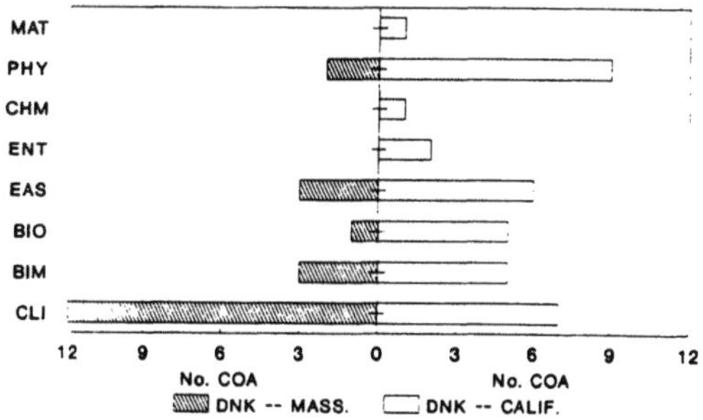


CNRS-LEPI MEV-MAC

## 6. Discussion

International cooperation is the result of a consensus between two or more scientists. A collaborative project at the laboratory-to-laboratory level is a form of collective recognition of the value of the research. The accumulation of the output of such research – a quantitative data set of coauthored articles – reflects the activities of science in the international community.

In studying relationships between countries and research institutions using the bibliometric data derived from ISI's *Science Citation Index*, we find several aspects that require further discussion.

First, there is a fundamental problem pertinent to the data bank. It must be noted that what this study indicates is based on the *Science Citation Index* and, therefore, the results follow the structure of this particular data bank. SCI is a selective data base; in order to be included in SCI, a journal must meet a selection standard.[6] When publishing the outcome of international collaborative research, scientists tend to select a journal recognized in all the collaborators' countries, although not necessarily edited in one of their own countries. Therefore, there is a high probability that the scientists will choose to publish their results in an influential international journal on the SCI list. Nevertheless, this list is limited to publications considered to be the results of "mainstream" research.[7] In other words, research that is not reported in the world's most central journals is ignored, and coverage of small countries could be biased, as many collaboration results circulate through regional channels.[8] It is, therefore, always desirable to use more than one data bank. However, SCI is almost unique in that it includes multiple addresses of authors so that international coauthorship can be identified. A comparative study of different data banks is, thus, very difficult.

The biases of coverage of articles in SCI are mentioned for certain countries, for example, due to the languages or other problems. However, when publishing research results of international collaboration, scientists usually use common languages. Therefore, language bias for certain countries is not as pronounced in international collaboration as it would be in the comparison of the levels of productivity among such countries.

Second, the classification of the eight scientific fields must be reconsidered. We used Carpenter's classification, but some journals could be re-examined. Pharmaceutical research, for example, is now mostly classified in clinical medicine, but this choice is open to discussion. In addition, the classification of eight fields may be too imprecise. Utilization of subfields should be studied, and social sciences and humanities should be included.

Third, data banks have been developed for information-retrieval purposes, and therefore their utilization for science-policy studies requires proper knowledge of their principles of construction and indexing. When studying international cooperation, the problem of on-leave scientists is often discussed, because each publication is attributed to a country (and research institution) according to the affiliation of scientists and not by the nationality of each individual researcher. A paper written by two Greek scientists could be counted as an international coauthorship between the United States and Greece, if one of the scientists is on leave to the United States. This example suggests that when studying international collaboration, the term "collaboration" must be clearly defined. It also suggests that numbers of COAs and COPs obtained by affiliation of authors cover overall

---

[6] E. Garfield, *Citation Indexing – Its Theory and Application in Science, Technology and Humanities* (New York: John Wiley & Sons, 1979).

[7] J.D. Frame, "Mainstream Research in Latin America and the Caribbean," *Interciencia* 2 (1977): 143-148.

[8] J. Gaillard, "La science du tiers monde est-elle visible?" *La Recherche* 20 (1989): 636-640.

aspects of cooperation, but do not specify certain details. The interpretation of numbers, therefore, must take into account different elements that may influence the numbers. The interpretation of data must consult the details of the collaborative project itself: by examining the articles and, when possible, talking to the scientists who are doing the research.

With the appearance of CD-ROM, technical possibilities are rapidly changing. This technology makes it possible to focus more on the collaborative projects at a micro level, using more detailed and vast data. This is a period of transition in bibliometric studies: from a macro vision to a micro vision.

The internationalization of science is accelerating in universities in most countries; large countries in science have a lower proportion of internationally coauthored papers in their total production than do smaller countries because scientists can more easily find equipment, methodologies, and concepts in their own countries. Nevertheless, in higher education and research in general, personal contacts and visits abroad are becoming more and more indispensable. Trends show growing international scientific activity around the world. International programs are increasing in size and number, but the personal needs of scientists are always the driving force of international cooperation in science.

## Acknowledgements

I would like to thank Audrey Buyrn and Yoshiko Okubo for valuable suggestions and discussions related to this work.

The views expressed in this study are those of the author. They do not necessarily represent those of the OTA.

# NATIONAL SECURITY INFORMATION CONTROLS IN THE UNITED STATES: IMPLICATIONS FOR INTERNATIONAL ACADEMIC SCIENCE AND TECHNOLOGY

JOHN SHATTUCK
*Vice President for Government, Community and Public Affairs*
*Harvard University*
*79 John F. Kennedy Street*
*Cambridge, MA 02138*

## Abstract

During the past decade, the U.S. government expanded its definition of national security interests and consequently felt a need to control the flow of scientific information in ways that may imperil its availability, shape its content, and limit its communication. Censorship of publications by government employees, restriction of unclassified scientific papers, curtailment of data collection and dissemination by federal agencies, and surveillance of computer network and library users are some of the controls that were introduced during this period. Recently, there have been indications that the trend is reversing and that less restrictive policies are in the process of being adopted. But much remains to be done before the system of information controls is sufficiently scaled back.

## Introduction

Scientific information is a critical resource in our modern technological world. The management of this resource, especially by government, has far-reaching implications for the quality and character of advanced industrial societies.

In the United States, scientific and technical information has been managed by the federal government for more than a decade in ways that have begun to restrict its availability, shape its content and limit its communication in a growing number of areas. In some respects this trend is predictable: resource management is an important function of government, and scientific and technical information is no less essential to national well-being than transportation systems, natural resources, or commercial markets. Nevertheless, government management of information in a democracy should be guided by a heavy presumption that openness of communication has great social utility. This presumption should be overridden only in the case of a substantial public necessity, and it should delineate the fine line between management and censorship of information.

A broad category of public necessity – the need to protect national security – has been advanced in the United States as a justification for overcoming the presumption against government control of the flow of scientific information. Until the late 1970s, the



*D. S. Zinberg (ed.), The Changing University*, 153–163.

prevailing philosophy in this area was expressed by President Truman's science adviser, Vannevar Bush, in 1945: "[a] sounder foundation for our national security rests in a broad dissemination of scientific knowledge upon which further advances can more readily be made than in a policy of restriction which would impede our further advances in the hope that our potential enemies will not catch up with us."[1] In the past decade this philosophy has been eroded by a policy of restricting the communication of broad categories of information on the ground that it must be kept from hostile ears and eyes.

The cumulative impact of this policy has had a restraining effect on academic inquiry, scientific and technological progress, economic activity, and democratic decision making. In some areas the policy has resulted in the suppression of the exchange of information to which the government attaches security importance, among scholars and researchers. In other areas the policy has restricted public access to the results of publicly funded academic research and other information-gathering activities. In still other areas it has deprived government officials of the technical and analytical bases for making informed decisions about how to carry out their functions.

## Information Policy as an Instrument of National Security Protection

For more than a decade, the defense and intelligence branches of the U.S. government have played an increasingly important role in shaping public information policies, influencing a wide variety of communications between the public and private sectors. During this period many government officials have come to view federal scientific and technical information policy as an instrument of national security protection.

The concept of national security itself underwent an apparently limitless expansion in the United States during the 1980s. Richard V. Allen, a former national security adviser to President Reagan, explained the concept in a 1983 report:

> In the 1980's "national security" is itself an all-encompassing term too often narrowly construed as having to do only with foreign policy and defense matters. In reality, it must include virtually every facet of international activity, including (but not limited to) foreign affairs, defense, intelligence, research and development policy, outer space, international economic and trade policy, and reaching deeply into the domains of the Departments of Commerce and Agriculture. In a word, "national security" must reflect the presidential perspective, of which diplomacy is but a single component.[2]

---

[1] Vannevar Bush, *Science - The Endless Frontier* (Washington, D.C.: U.S. Government Printing Office, 1945), p.182.

[2] Richard V. Allen, "Foreign Policy and National Security: The White House Perspective," in *A Mandate for Leadership Report*, edited by R.N. Holwill (Washington, D.C.: The Heritage Foundation, 1983), p.6.

Reflecting this panoramic definition, a broad national security net has been placed around the open communication of much scientific, technical, economic and political information. The net includes an expanded classification system, the use of export controls to restrict the flow of technical data, limits on the communication of so-called sensitive unclassified information and additional restraints on communication between U.S. citizens and foreigners.

The asserted justification for these restrictive policies is that certain categories of information must be kept from hostile foreign governments, international competitors, terrorists and others who would damage the national security by misusing the information. At the center of this justification is a theory that some data are inherently dangerous and must be officially restricted – even if the government does not own or control them, if the data are already publicly available, if they are the product of a private discovery, or if they are not in themselves likely to cause damage if disseminated. This is the theory of an information mosaic – bits and pieces of seemingly harmless information that can be assembled through sophisticated search techniques in such a way as to be harmful in the aggregate.[3] An often-cited example is the blueprint for manufacturing an H-bomb that was drawn from unclassified information scattered through open scientific journals and published in a 1979 article in *The Progressive* magazine.[4]

### *A. Expanded Classification and Academic Research*

A major policy achievement of proponents of the mosaic theory of information has been a significant expansion of the security classification system for keeping secret certain categories of scientific and technical data.

The Carter administration's executive order on classification,[5] replaced by President Reagan in 1982,[6] contained a variety of features that favored disclosure over secrecy. The earlier order stipulated that even if information met one of seven classification categories,[7] it was not to be classified unless "its unauthorized disclosure reasonably could be expected to cause at least identifiable damage to the national security."[8]

---

[3] See, e.g., Richard Perle, "The Eastward Technology Flow: A Plan of Common Action," *Strategic Review* (Spring 1984): 24.

[4] Howard Morland, "The H-bomb Secret," *The Progressive* (November 1979): 14.

[5] Executive Order 12065 (1978).

[6] Executive Order 12356 (1982).

[7] The categories were: "a) military plans, weapons or operations; b) foreign government information; c) intelligence activities, sources or methods; d) foreign relations or foreign activities of the U.S.; e) scientific, technological or economic matters relating to national security; f) United States government programs for safeguarding nuclear facilities; or g) other categories of information which...are related to national security and which require protection against unauthorized disclosure."

[8] Executive Order 12065, Sec. 1-302.

Requiring automatic declassification after six years,[9] the earlier order established a presumption that "[i]f there is a reasonable doubt" which designation is appropriate, or whether the information should be classified at all, the less restrictive designation should be used, or the information should not be classified.[10]

The government's 1982 overhaul of classification policies created new procedures to limit the availability of information. It established a presumption in favor of classification in all cases in which officials are in doubt whether secrecy is necessary; it eliminated the requirement of automatic declassification of information within a prescribed length of time; and it created new authority of officials to reclassify information that is already in the public domain.[11]

The expanded classification system gives authority to government officials to impose classification restrictions over federally funded research projects after the contracts have been signed and the projects undertaken. The order appears to allow classification to occur at any stage of a research project and to be maintained indefinitely.[12] The net effect could be to inhibit researchers from making long-term intellectual investments in nonclassified projects in such fields as cryptography or laser science that have features that make them likely subjects for classification at a later date.

The publication of federally funded research is also affected by the expanded classification system. Publication decisions are governed by National Security Decision Directive 189, promulgated in response to concerns that unclassified research sponsored by the federal government might be restricted.[13] The directive provides that restrictions on publication can result only from classification:

> It is the policy of this Administration that, where the national security requires control, the mechanism for control of information generated during federally-funded fundamental research in science, technology and engineering at colleges, universities and laboratories is classification.[14]

Under the 1982 classification order, however, federal officials are authorized to reclassify information that is in the public domain if the information "may reasonably be recovered."[15] An example of how this reclassification policy can affect federally

---

[9] Ibid, Sec. 1; 1-4 Duration of Classification. There is one exception to this rule: foreign-government information may be classified up to thirty years.

[10] Ibid, Sec. 1; 1-1 Classification Designation, 1-101.

[11] Executive Order 12356 (1982). See Glenn English, "Congressional Oversight of Security Classification Policy," *Government Information Quarterly*, 2 (1984): 166.

[12] Executive Order 12356, Sec. 1.6(d).

[13] "National Policy and Transfer of Scientific, Technical and Engineering Information," September 21, 1985.

[14] Ibid., p. 2.

[15] Executive Order 12356, Sec. 1.6(b).

sponsored research is the following language from a research grant agreement with the Department of Energy:

> It is not expected that activities under this grant will generate or otherwise involve classified information (i.e., Restricted Data, Formerly Restricted Data, National Security Information). However, if in the opinion of the grantee or DOE such involvement becomes expected prior to the closeout of the grant, the grantee or DOE shall notify the other in writing immediately. If the grantee believes any information developed or acquired may be classifiable information to anyone, including DOE officials with whom the grantee normally communicates, except the Director of Classification, it shall protect such information as if it were classified until notified by DOE that a determination has been made that it does not require such handling...[16]

This language specifies no limits on the length of time a researcher is obligated to protect "potentially classifiable" information. Thus, some scholars fear that under the current classification order, "[a]cademic research not born classified may...die classified."[17]

*B. Restrictions on the Communication of Science and Technology*

The expanded classification system reaches information collected by or under the sponsorship of the federal government. As we have seen, however, the prevailing view of U.S. national-security interests sweeps beyond the government's own information activities, calling for controls over categories of civilian data as well. In recent years such controls have been developed in a growing number of scientific and technological fields.

The asserted basis for restricting scientific and technical data is that they are different from other kinds of information that can be freely communicated. First, technical data can be used to create things that are intrinsically dangerous, such as weapons systems. Second, such data can have an immediate economic utility and are thus more like commodities than ideas. Third, scientific discovery often results from the direct involvement of government as a sponsor of research. Based on these justifications, an extensive system of export controls over categories of technical data has been devised; new controls have been developed over the kinds of communications scientists can have among themselves; and additional limitations on the communication of "sensitive unclassified" information have been proposed.

Export control laws in the United States were originally enacted to regulate the overseas export of goods and machinery. During the 1980s, however, they were increasingly applied to restrict the communication of technical information and ideas

---

[16] Document on file with Harvard University, Office for Sponsored Research.

[17] Robert A. Rosenbaum, Morton J. Tenzer, Stephen H. Unger, William Van Alstyne, Jonathan Knight, "Academic Freedom and the Classified Information System," *Science* 219 (January 21, 1983): 257.

within the United States. An example of this application involved efforts by the Departments of Defense and Commerce to persuade scientific and engineering societies to limit access to professional conferences at which unclassified technical papers are presented. Throughout the decade, the proceedings of several professional societies were affected by export control policies, including the American Vacuum Society,[18] the Institute of Electrical and Electronic Engineers,[19] the American Ceramics Society,[20] the Institute of Aeronautics and Astronautics,[21] the Society for the Advancement of Material and Process Engineering,[22] the Society of Photo-Optical Instrumentation Engineers,[23] and the Linear Accelerator Conference.[24]

In order to avoid problems of clearance, some scientific and technical societies informally barred foreign scientists from attending their meetings. These included such prestigious associations as the Society of Manufacturing Engineers, the American Ceramics Society and the Society for the Advancement of Material and Process Engineering. Nevertheless, these were the exceptions. According to a 1986 survey conducted by the American Association for the Advancement of Science, "in cases where societies have adopted policies regarding foreign participation in their meetings, two-thirds of the societies prohibit society sponsorship of closed or restricted sessions."[25]

Under the export control system, classrooms, libraries and research laboratories are all areas where U.S. scientists have had to exercise caution about foreign contacts. For example, a materials science course offered at the University of California at Los Angeles in 1984, "Metal Matrix Composites," initially had to be limited to students who were U.S. citizens because the course involved unclassified technical data appearing on an export control list.[26] Another example was an effort by the Department of State in 1981 to require universities to report campus contacts between U.S. citizens and Chinese exchange students.[27] Similary, in 1987 the Federal Bureau of Investigation initiated a "Library Awareness Program" within its counterintelligence effort to urge librarians to identify

---

[18] Nicholas Wade, "Science Meetings Catch the U.S. Soviet Chill, *Science* 207 (March 7, 1980): 1056-1058.

[19] Nicholas Wade, "Government Bars Soviets from AVS and OSA Meetings," *Physics Today* (April 1980): 81-83.

[20] Robert L. Park, "Science and Secrecy," *Bulletin of the Atomic Scientist* (March 1985): 22.

[21] Ibid, p. 23.

[22] Ibid, p. 23.

[23] Ibid, p. 22.

[24] Memorandum from Robert L. Park, American Physical Society, September 4, 1986.

[25] "Access to Scientific and Technical Information," *American Association for the Advancement of Science Bulletin* (Summer 1986); "Censorship and Self-Censorship of Scientific Communication," paper by Robert L. Park, The American Physical Society, January 1985.

[26] M. Wallerstein and L. McCray, "Update of the Corson Report," January 26, 1984 (unpublished staff report for National Academy to Sciences), p. 9.

[27] National Academy of Sciences, *Sciences Communication and National Security* (Washington, D.C.: National Academy Press, October 1982), p. 172-181. The study is also known as the Corson Report.

library users who might be "hostile intelligence people."[28] Finally, the Department of Defense for a time required universities engaged in DOD-sponsored research in artificial intelligence to certify that persons who receive technical data generated by the research are U.S. or Canadian citizens.[29]

Another area in which U.S. scientists have had difficulty collaborating with their foreign counterparts involves scientific instrumentation and equipment. A prominent example is the use of supercomputers – the next generation of computing technology. Supercomputers have computational capacities many times greater than other computers. The National Science Foundation has provided a major portion of the funding for five supercomputers at university consortium centers with the intention that they will be used for unclassified basic research. Proposed federal guidelines to limit foreign scholars' access to supercomputers drew expressions of dismay in 1986 from research scientists, who feared that restraints on the use of unclassified facilities would be harmful to the quality of academic science.[30] After much controversy, the guidelines were revised.

Beyond these restrictions on contacts between U.S. and foreign scientists, the U.S. government has designated general categories of scientific and technical research as inherently sensitive and therefore subject to information control. A prominent example is cryptography. Since 1981 that field has been affected by a National Security Agency designation as sensitive, and many cryptologists now voluntarily submit their work to NSA for prepublication reviews in order to avoid running afoul of the export-control laws.[31] Nuclear energy is another scientific field that is increasingly secret. In 1981 Congress amended the Atomic Energy Act to authorize the Secretary of Energy to regulate "the unauthorized dissemination of unclassified nuclear information."[32]

A final area of scientific and technical information targeted until recently for U.S. government control involves information in electronic data bases. This is by far the largest category of potentially restricted information, because it can be found in any academic, commercial or governmental computerized information system. The theory behind the need to control this information is the information mosaic – bits of seemingly harmless data that can be assembled through sophisticated electronic searching in such a way as to be damaging in the aggregate.

Much of the attention in this area was focused on a National Security Council directive

---

[28] Robert D. McFadden, "Libraries are asked by FBI to Report on Foreign Agents," the *New York Times*, September 18, 1987.

[29] Letter from Donald K. Hess, Vice President for Administration, University of Rochester to the author, November 6, 1987, and attached DOD Form 2345.

[30] Letter from Richard G. Leahy, Chairman, Board of Trustees, Consortium for Scientific Computing, to William Schneider, Jr., Undersecretary of State for Security Assistance, Science and Technology, May 8, 1986.

[31] Kolata, "NSA Asks to Review Papers Before Publication," *Science* 215 (March 19, 1982): 1485, and "Prior Restraints on Cryptography Considered," *Science* 208 (June 27, 1980): 1442-1443.

[32] 42 U.S.C. 2168(a)(1).

promulgated in October 1986 by John Poindexter, President Reagan's national security adviser.[33] The Poindexter directive sought to restrict not only unclassified information affecting national-security interests, but also any computerized information that could adversely affect "other government interests," including but not limited to "government or government-derived economic, human, financial, industrial, agricultural, technological, and law-enforcement information."[34]

Poindexter's directive raised the specter of U.S. intelligence agencies monitoring and regulating virtually all academic and commercial computerized data bases and information exchanges in the United States. The directive was withdrawn in March 1987 under congressional pressure, but the underlying policy remained in place.[35] The policy was set out in National Security Decision Directive 145, which calls for a "comprehensive and coordinated approach" for all telecommunications and automated information systems, under the theory that "information, even if unclassified in isolation, often can reveal sensitive information when taken in the aggregate.[36]

Recently, the Bush administration announced that it would begin scaling down the Reagan administration's effort to restict the flow of computerized information. As a first step it began disbanding the National Computer Security Center, which was established in 1981 to develop strategies for monitoring database communications.

## Long-Term Impact of National Security Controls Over Scientific Communication

What are the likely long-term effects of a national security policy that restricts the communication of science and technology?

The first and most obvious is that too many restrictions can lead to a stagnation of basic science. The Soviet Union offers an example of what can happen. According to the American Physical Society, Soviet research and development in the fields of solid-state electronics and biology have been severely retarded because of pre-*glasnost* official restrictions on scientific communication.[37] The Defense Department reported in 1987 that in 20 key technologies, the United States was leading the Soviet Union in 14 and was

---

[33] National Telecommunications and Information Systems Security Policy No. 2, "National Policy on Protection of Sensitive, but Unclassified Information in Federal Government Telecommunications and Automated Information Systems," October 29, 1986.

[34] Ibid., p. 1.

[35] Letter from Howard H. Baker, Chief of Staff to the President, to the Hon. Jack Brooks, March 16, 1987.

[36] National Security Decision Directive 145, "National Policy on Telecommunications and Automated Information Systems Security," September 17, 1984.

[37] Testimony of American Physical Society, before a hearing of the Committee on Government Operations, U.S. House of Representatives, February 25, 1987.

at least tied in the other six.[38] More generally, the National Academy of Sciences concluded in a 1982 report that the continued health of U.S. science depends heavily on open communication.[39]

A second negative effect of the current system of technical information controls is on the economy. Another National Academy of Sciences report issued in April 1987 estimated that the cost of the existing regime of export controls was 188,000 jobs and $9 billion a year.[40] Since the U.S. economy depends substantially on foreign technical expertise, a strict regime of export controls will have a negative domestic economic impact because it will limit the employment of needed technical experts. One unintended result of export controls is the funneling of business to foreign competitors. According to the NAS Report, 38 percent of U.S. exporting companies have reported losses of sales to foreign competitors because of export controls.[41] Another economic problem is the inability of U.S. investors to get information about new corporate technologies that are covered by such controls.

A third negative effect, ironically, is likely to be on U.S. national security itself. Most experts agree that the long-term security needs of the United States depend on rapid technological development, which is not possible if broad communication restrictions are in place.[42] Last but certainly not least, democratic values, freedom of speech, and the openness of U.S. society are likely to be eroded if the current policy trend continues.

All these damaging effects of excessive secrecy in the area of science and technology can be illustrated by a widely-publicized example of some of the benefits of open scientific communication that could be lost under a restrictive information policy. In mid-March 1987 thousands of physicists from around the world gathered in New York at a meeting of the American Physical Society to exchange information about the latest developments in the discovery of superconductors that conduct the flow of electricity without losing energy at significantly higher temperatures than previously possible.[43] Most scientists believe these new materials will revolutionize a whole range of existing technologies – from electrical power generation and transmission to computers and telecommunications.

The story began in 1986 at an IBM lab in Zurich, where a Swiss and a German scientist succeeded in creating a relatively high-temperature superconducting ceramic

---

[38] Ibid.

[39] National Academy of Sciences, *Scientific Communication and National Security* (Washington, D.C.: National Academy Press, 1982).

[40] National Academy of Sciences, *Balancing the National Interest: U.S. National Security, Export Controls and Global Economic Competition* (Washington, D.C.: National Academy Press, 1987).

[41] Ibid.

[42] See e.g., Edward Teller, "Secrecy in Science," *Los Angeles Times Magazine*, July 5, 1987, p. 4B; Harold Willenbock, "Information Controls and Technological Progress," *Issues in Science and Technology* (Fall 1986): 88.

[43] James Gleick, "In the Trenches of Science, *The New York Times Magazine*, August 16, 1987, p. 77.

material. The discovery was published in an international scientific journal, and scientists at universities and corporate research laboratories in Houston, Tokyo and other cities began a race to develop a practical method of raising the temperature for superconductors so that the revolutionary new technology could be put to use. Many of the physicists engaged in the race were not U.S. citizens.[44]

Only one segment of the industrialized world was left out in the cold during this extraordinarily fertile period of discovery and communication. Not surprisingly, perhaps, pre-*glasnost* Warsaw Pact nations played no part in the superconductor race. No one sought to exclude them, but they were weighted down with travel restrictions, restraints on contacts with foreigners, and widespread suspicion of telephones and copying machines.[45] Ironically, the United States showed signs of emulating some of these restrictions by holding a White House conference on superconductivity in July 1987 that excluded foreign science officials.[46]

## Conclusion

With the superconductor experience as background, the outlook is improving for a relaxation of restrictions on the communication of science and technology that have been imposed in the U.S. during the past decade.

There are at least three reasons why the public is growing skeptical about the need for broad information controls. First, the extraordinary changes in Eastern Europe and the Soviet Union have called into question some aspects of the underlying security premises of a restrictive information policy. Second, national concern about U.S. competitiveness has created strong pressures for unleashing science and technology to better serve the economy. Third, after a decade of growing secrecy there is little evidence that restrictive scientific and technical information policies have achieved their national-security purposes.

At the international level there are substantial pressures on the U.S. to adopt a less restrictive policy on scientific and technical communication. Many of the technologies that the current regime of controls attempts to reach are increasingly uncontrollable in a global economy driven by rapid technological change. This is certainly the perspective of Western European governments, and it was forcefully articulated by the National Academy of Sciences in its 1988 report, *Global Trends in Computer Technology and Their Impact on Export Controls*.[47]

On the other hand, the changes that are transforming the political map of the world

---

[44] Ibid.

[45] Ibid.

[46] Michael Specter, "Superconductivity Conference to Exclude Foreign Officials," the *Washington Post*, July 25, 1987, p. A2.

[47] National Academy of Sciences, *Global Trends in Computer Technology and Their Impact on Export Controls* (Washington, D.C.: National Academy Press, 1988).

today are creating great uncertainty, instability and anxiety, which could retard efforts to make fundamental changes in restrictive information policies. Instability in the Soviet Union, Eastern Europe, and the Middle East, international competition for technological markets, pressures for economic protectionism, and concerns about terrorism in unstable parts of the world may combine to slow the movement toward open scientific communication and the globalization of research. These negative factors can be moderated by international collaborative efforts in science and engineering. Research projects can be promoted involving international teams of academic scientists; national policies on scientific communication can be included on the common defense agenda of NATO; the network of research university exchanges across national borders can be improved; the international demand for technological advances in such areas as superconductivity can be used to dramatize the importance of reducing restrictions on scientific and technical communication; and research universities can be more vigilant in their relations with government to ensure that agreements for sponsored research do not impose unacceptable conditions on publishing the results.

Above all, the international scientific community can rededicate itself to the proposition that the free flow of information and ideas is vital to academic science and technology. The engines of innovation that drive economic growth and guarantee national security are powered by open and unfettered communication. A lesson suggested by the U.S. experience with broad national security information controls over the past decade is that such restrictive policies are counterproductive and ultimately self-defeating.

# LIST OF CONTRIBUTORS

Dr. John A. Armstrong
Vice President for Science and Technology
IBM Corporation
Old Orchard Road
Armonk, NY 10504
U.S.A

Mr. Erich Bloch
Director, National Science Foundation (1984-90)
2801 New Mexico Avenue
Washington, DC 20007
U.S.A.

Mr. Derek C. Bok
President
Harvard University
Massachusetts Hall
Cambridge, MA 02138
U.S.A.

Professor Harvey Brooks
Benjamin Peirce Professor of Technology and Public Policy Professor of Applied Physics on the Gordon Mckay Endowment, Emeritus
John F. Kennedy School of Government
79 John F. Kennedy Street
Cambridge, MA 02138
U.S.A.

Dr. George Bugliarello
President, New York Polytechnic University
33 Jay Street
Brooklyn, NY 11201
U.S.A.

Mr. Alan Fechter
Director
Office of Scientific and Engineering Personnel
Green Building, Room 415
2101 Constitution Avenue, N.W.
Washington, D.C. 20418
U.S.A.

Professor Fumio Kodama
Director, Research Group, National Institute of Science and Technology Policy
Science and Technology Agency
1-11-39, Nagata-Cho, Chiyoda ku
Tokyo 100,
Japan

Dr. Jean-Francois Miquel
Director, LEPI/CNRS U.S. Congress
Centre National de la Recherche Scientifique
Laboratory for Evaluation and Prospects of International Scientific Activities
CNRS
295 rue St. Jacques
7500 Paris
France

Mr. John Shattuck
Vice President for Government, Community and Public Affairs
Harvard University
Massachusetts Hall
Cambridge, MA 02138
U.S.A.

Professor Shirley Williams
Public Service Professor of Electoral Politics
John F. Kennedy School of Government
Harvard University
79 John F. Kennedy Street
Cambridge, MA 02138
U.S.A. (from the U.K.)

Ms. Maureen Woodhall
Institute of Higher Education
University of London
58/59 Gordon Square
London WC1
U.K.

Dr. Dorothy S. Zinberg
Lecturer in Public Policy, Science, Technology & Public Policy Program
Senior Research Fellow, Center for Science and International Affairs
Kennedy School of Government
Harvard University
79 John F. Kennedy Street
Cambridge, MA 02138
U.S.A.

# LIST OF PARTICIPANTS

Professor Armando Albert
Vice President
Consejo Superior de Investigaciones Cientificas - CSIC
Calle Serrano 17
E-28006
Madrid
Spain

Dr. John A. Armstrong
Vice President for Science and Technology
IBM Corporation
Old Orchard Road
Armonk, NY 10504
U.S.A.

Dr. Elinor Barber
Low Library #310
Columbia University
New York, NY 10027
U.S.A.

Mr. Erich Bloch
Director
National Science Foundation (1984-90)
2801 New Mexico Avenue
Washington, DC 20007
U.S.A.

Mr. Derek C. Bok
President
Harvard University
Massachusetts Hall
Cambridge, MA 02138
U.S.A.

Ms. Jennifer Bond
Science Resources, National Science Foundation
1800 G Street, N.W.
Washington, D.C. 20550
U.S.A.

Professor Lewis M. Branscomb
Albert Pratt Public Service Professor
Director, Program for Science, Technology and Public Policy
John F. Kennedy School of Government
Harvard University
79 John F. Kennedy Street
Cambridge, MA 02138
U.S.A.

Professor Harvey Brooks
Benjamin Peirce Professor of Technology and Public Policy
Professor of Applied Physics on the Gordon Mckay Endowment, Emeritus
John F. Kennedy School of Government
Harvard University
79 John F. Kennedy Street
Cambridge, MA 02138
U.S.A.

Dr. George Bugliarello
President
New York Polytechnic University
33 Jay Street
Brooklyn, NY 11201
U.S.A.

Professor Ashton Carter
Professor of Public Policy
Associate Director, Center for Science and International Affairs
John F. Kennedy School of Government
Harvard University
79 John F. Kennedy Street
Cambridge, MA 02138
U.S.A.

Professor Paul Doty
Mallinckrodt Professor of Biochemistry
Director Emeritus, Center for Science and International Affairs
John F. Kennedy School of Government
Harvard University
79 John F. Kennedy Street
Cambridge, MA 02138
U.S.A.

Dr. Jacques Ducuing
Director
Scientific Affairs Division
North Atlantic Treaty Organization (NATO)
B-1110 Brussels
Belgium

Dr. Gerald Epstein
Dual Use Project Director
Science, Technology and Public Policy Program
John F. Kennedy School of Government
Harvard University
79 John F. Kennedy Street
Cambridge, MA 02138
U.S.A.

Mr. Alan Fechter
Director
Office of Scientific and Engineering Personnel
Green Building, Room 415
2101 Constitution Avenue, N.W.
Washington, D.C. 20418
U.S.A.

Dr. Rolf Hoffmann
Alexander von Humboldt-Stiftung (AvH)
Jean-Paul Strasse 12
D-5300 Bonn 2
F.R.G.

Dr. Barry Horowitz
Executive Vice President
The MITRE Corporation
Burlington Road
Bedford MA 01730
U.S.A.

Dr. John C. Hoy
President
New England Board of Higher Education
45 Temple Place
Boston, MA 02111
U.S.A.

Dr. Keneichiro Imai
Member, Science Congress
Minami-Aoyama 1-26-6
Minato-ku
Tokyo 107
Japan

Professor Fumio Kodama
Director, Research Group
National Institute of Science and Technology Policy
Science and Technology Agency
1-11-39, Nagata-Cho, Chiyoda ku
Tokyo 100
Japan

Dr. George Leitmann
Professor of Engineering Sciences
Associate Dean of Academic Affairs
University of California
Berkeley, CA 94740
U.S.A.

Dr. Roberta Balsted Miller
Director, Division of Social and Economic Science
National Science Foundation
1800 G Street, NW
Washington, DC 20550
U.S.A.

Dr. Hassan Minor
President, The Partnership
315 Commonwealth Avenue
Boston, MA 02115
U.S.A.

Dr. Jean-Francois Miquel
Director, LEPI/CNRS
U.S. Congress
Centre National de la Recherche Scientifique CNRS
Laboratory for Evaluation and Prospects of International Scientific Activities
295 rue St. Jacques
7500 Paris
France

Professor Pedro Pascual
Asesor Cientifico del Gabinete de la Secretario de Estado
de Universidades e Investigacion
Serrano 150
E-28006 Madrid
Spain

Dr. Don I. Phillips
Executive Director
Government, University and Industry Research Roundtable
National Academy of Sciences
2101 Constitution Ave, NW
Washington, DC 20418
U.S.A.

Mr. John Shattuck
Vice President for Government, Community and Public Affairs
Harvard University
Massachusetts Hall
Cambridge, MA 02138
U.S.A.

Dr. Klaus Shroeter
Science Counselor
Embassy of the Federal Republic of Germany
4645 Reservoir Street
Washington, DC 20007 U.S.A.
(from F.R.G.)

Dr. Manfred Stassen
American Institute for Contemporary German Studies
11 Dupont Circle, NW
Suite 350
Washington, DC 20036
U.S.A.

Representative Robert G. Torricelli
U.S. House of Representatives
317 Cannon HOB
Washington, DC 20515-3009
U.S.A.

Dr. E. van Spiegel
Director-General for Science Policy
Ministry of Education and Science
Europaweg 4
P.O. Box 25000
2700 LZ Zoetermeer
The Netherlands

Professor Shirley Williams
Public Service Professor of Electoral Politics
John F. Kennedy School of Government
Harvard University
79 John F. Kennedy Street
Cambridge, MA 02138 U.S.A.
(from the U.K.)

Dr. Linda Wilson
President
Radcliffe College
Harvard University
Cambridge, MA 02138

Ms. Maureen Woodhall
Institute of Higher Education
University of London
58/59 Gordon Square
London WC1
U.K.

Dr. Nicolas Ziegler
Visiting Assistant Professor of International Manangement
Sloan School of Management
Massachusetts Institute of Technology
E52-546
Cambridge, MA 02139
U.S.A.

Professor John Ziman
Director
Science Policy Support Group
22, Henrietta Street
London WC2E 8NA
U.S.A.

Dr. Dorothy S. Zinberg
Lecturer in Public Policy, Science, Technology & Public Policy Program
Senior Research Fellow, Center for Science and International Affairs
Kennedy School of Government
Harvard University
79 John F. Kennedy Street
Cambridge, MA 02138
U.S.A.

Mr. Charles Zraket
Chief Executive Officer
The MITRE Corporation
Burlington Road
Bedford, MA 01730
U.S.A.

# INDEX

MIX
Papier aus verantwortungsvollen Quellen
Paper from responsible sources
FSC® C105338

If you have any concerns about our products,
you can contact us on
**ProductSafety@springernature.com**

In case Publisher is established outside the EU,
the EU authorized representative is:
**Springer Nature Customer Service Center GmbH**
**Europaplatz 3, 69115 Heidelberg, Germany**

Printed by Libri Plureos GmbH
in Hamburg, Germany